Urban Sustainability

Editor-in-Chief

The Urban Sustainability Book Series is a valuable resource for sustainability and urban-related education and research. It offers an inter-disciplinary platform covering all four areas of practice, policy, education, research, and their nexus. The publications in this series are related to critical areas of sustainability, urban studies, planning, and urban geography.

This book series aims to put together cutting-edge research findings linked to the overarching field of urban sustainability. The scope and nature of the topic are broad and interdisciplinary and bring together various associated disciplines from sustainable development, environmental sciences, urbanism, etc. With many advanced research findings in the field, there is a need to put together various discussions and contributions on specific sustainability fields, covering a good range of topics on sustainable development, sustainable urbanism, and urban sustainability. Despite the broad range of issues, we note the importance of practical and policy-oriented directions, extending the literature and directions and pathways towards achieving urban sustainability.

The series will appeal to urbanists, geographers, planners, engineers, architects, governmental authorities, policymakers, researchers of all levels, and to all of those interested in a wide-ranging overview of urban sustainability and its associated fields. The series includes monographs and edited volumes, covering a range of topics under the urban sustainability topic, which can also be used for teaching materials.

Fangqu Niu

Urban Spatial Evolution Simulation

Theory, Method, and Practice

中国工信出版集团 電子工業出版社 PUBLISHING HOUSE OF ELECTRONICS INDUSTRY http://www.phei.com.cn

Springer

Fangqu Niu
Institute of Geographic Sciences
and Natural Resources Research
Chinese Academy of Sciences
Beijing, China

ISSN 2731-6483 ISSN 2731-6491 (electronic)
Urban Sustainability
ISBN 978-981-97-3480-1 ISBN 978-981-97-3481-8 (eBook)
https://doi.org/10.1007/978-981-97-3481-8

Jointly published with Publishing House of Electronics Industry Co., Ltd
The print edition is not for sale in China (Mainland). Customers from China (Mainland) please order the print book from: Publishing House of Electronics Industry Co., Ltd.

This work was supported by National Natural Science Foundation of China (42071153, 41101119).

Translation from the Chinese Simplified language edition: "城市空间演化模拟理论、方法与实践" by Fangqu Niu, © Publishing House of Electronics Industry Co., Ltd 2023. Published by Publishing House of Electronics Industry Co., Ltd. All Rights Reserved.

This Springer imprint is published by the registered company Springer Nature Singapore Pte Ltd.
The registered company address is: 152 Beach Road, #21-01/04 Gateway East, Singapore 189721, Singapore

Preface

As the advancement of new urbanization continues, an increasing number of people will migrate to cities in China, leading to a series of transformations in urban economic activities, transportation, and spatial structure. Formulating scientific and rational urban space policies is essential for ensuring the healthy and orderly development of urban spaces. Urban spatial decision-making often needs to answer questions like, 'What impact would developing a certain amount of housing or commercial floorspace at a particular location have on urban space?' or 'What effects would constructing a new highway (or rail transit) have on the distribution of population and businesses?' Simulating the urban spatial evolution process and conducting policy experiments are vital for ensuring the reasonableness and effectiveness of urban spatial policies.

Urban spatial simulation research has a long history. Since the quantitative revolution in geography in the 1960s, models have been used internationally to study urban spatial evolution issues, such as the Lowry model. Since the 1990s, significant advancements have been made in micro-simulation methods based on artificial intelligence theory, where a bottom-up simulation approach studies the macroevolution laws of urban space by simulating the behavior of micro-agents. In summary, the commonly used models mainly include two categories: bottom-up models based on Cellular Automata (CA) technology and top-down Land Use-Transportation Interaction (LUTI) models.

Currently, CA models are widely studied domestically, focusing on the simulation of physical land use changes. The integration of CA models with agent technology further enhances their simulation capabilities. However, the application and research of LUTI models are still very limited, often remaining at the conceptual understanding stage, with few successful application cases. The LUTI model is considered a powerful tool for simulating the spatial evolution process of urban socio-economic activities and is often used for decision-making support in urban development in developed countries. Unlike the CA model, which simulates macro-urban space laws through micro-individuals, the LUTI model is constructed in a top-down manner, based on urban development principles, using macro-aggregated data to describe urban space. Understanding and comprehending the LUTI model is essentially understanding the principles of urban development. Therefore, learning the LUTI model is an important way to understand the process of urban spatial evolution and master its simulation methods. To promote the development and application of the LUTI model domestically, and aid the quantitative research of human geography in China, we have condensed and summarized recent research achievements to write this book.

This book primarily introduces the principles, development history, and modeling methods of the LUTI model, contributing to its promotion and application in China. It begins with a brief overview of cities, urbanization, urban space, and other related concepts, and summarizes the commonly used methods of urban spatial evolution simulation. On this basis, it clarifies the concept, origin, and developmental history of the LUTI model, proposes the LUTI model's modeling framework, and then details the implementation process of each component of the model. After systematically introducing the LUTI model, a field still underdeveloped in domestic research, the book establishes a LUTI model based on 'activity' modeling technology, named the ActSim model, to demonstrate the construction process of the LUTI model. The 'activity'-based modeling technology aligns with the development forefront of the LUTI model, and this book provides an important reference for domestic scholars to understand and apply the LUTI model. While enriching domestic urban simulation research methods, it also aids in the development and application of the LUTI model.

This book is suitable for graduate students in human geography and related fields and can also provide a reference for technical and research personnel in related areas of urban studies. The LUTI model is used to simulate the urban spatial evolution process, and urban development involves various aspects, making it infinitely complex. Although this book provides a framework for the 'activity'-based LUTI model and initially implements the model (ActSim model), the ActSim model still has many functions that urgently need to be improved, such as simulating family changes, different socio-economic groups, and transportation modes. We will continue to deepen our research and constantly improve the ActSim model and this book.

The book consists of 11 chapters. Chapter 1 briefly outlines concepts related to cities, urbanization, and urban space. Chapter 2 introduces the related theories, influencing factors, and research significance of urban spatial evolution. Chapter 3 provides a brief analysis of urban spatial analysis and simulation models. Chapters 4–9 illustrate the construction process of the LUTI model, with the built model simulating changes in urban activity locations, named the ActSim model. Chapter 4 presents the framework and structure of the ActSim model; Chap. 5 introduces the urban building development prediction model; Chaps. 6 and 7 respectively introduce urban location cost and transportation accessibility evaluation methods; Chap. 8 establishes the location model based on the content of previous chapters, used to predict the location of urban families and enterprises; Chap. 9 presents the case application of the ActSim model and discusses it in detail. As an extension of the urban model, Chap. 10 constructs a regional model framework to simulate the regional spatial evolution process, including frameworks for regional economic models and population migration models. As a supplement, Chap. 11 introduces commonly used development languages, analysis and simulation tools, and common functions.

I am responsible for writing the entire book, including setting the writing goals and content of the book, drafting the writing requirements and outlines for each chapter, and reviewing, proofreading, and approving the content of each chapter. Graduate students Bingcheng Xuan, Zhongling Xin, and Lijia Zhu participated in the collection and proofreading of materials for some chapters. Due to our limited abilities, there may be errors in the content of this book. Readers are welcome to correct them, and my email is niufq@lreis.ac.cn. I look forward to reader feedback.

Beijing, China Fangqu Niu
January 2024

Acknowledgments

The authors are grateful to David Simmonds Consultancy Limited, Cambridge, UK, for the permission to use DELTA package to implement the urban model. Many thanks to Prof. David Simmonds for his guidance on modeling. His talent and profound knowledge are admirable and have greatly benefited us.

Great thanks also go to graduate students Bingcheng Xuan, Zhongling Xin, and Lijia Zhu for their efforts in preparing the manuscript. Many thanks to Springer for their efforts in getting this book published.

I would also like to express my deepest gratitude to the National Natural Science Foundation of China (NSFC) for the significant financial support (42071153, 41101119).

Introduction

This book summarizes the theory and methods of urban spatial evolution simulation and demonstrates the research results using Beijing as a case study. It comprises 11 chapters, starting with a brief discussion on concepts such as cities, urbanization, and urban space, and elaborates on the commonly used methods in urban spatial evolution simulation. On this basis, it clarifies the concept, origin, and developmental history of the Land Use-Transportation Interaction (LUTI) model. Combining the principles of the LUTI model, the book interprets the process of urban spatial evolution, proposes a modeling framework for the LUTI model, and details the modeling process and case applications. As an extension of urban simulation, it further constructs a regional model framework to lay a foundation for subsequent regional simulation research. Lastly, the book provides a brief analysis of commonly used tools in urban spatial analysis and simulation. Systematically introducing the LUTI model, a field still underdeveloped in domestic research, and its advancements, this book serves as an important reference for domestic scholars to understand and apply the LUTI model, significantly enriching the methods of urban simulation research and promoting quantitative research in domestic human geography.

This book can serve as a reference for researchers in fields such as urban planning, geography, and economics. It can also be used as a textbook or reference book for senior undergraduates and graduate students in higher education institutions.

Contents

About the Author

Fangqu Niu, Ph.D. a visiting scholar at the University of Leeds, UK. He is currently an associate professor and special backbone personnel at the Institute of Geographic Sciences and Natural Resources Research, Chinese Academy of Sciences, and a supervisor for graduate students. His main research interests include urban and regional development simulation analysis, transportation geography, and the "Belt and Road Initiative". He innovatively developed an activity-based land use-transport interaction model taking Chinese cities as case studies. He has hosted or participated in more than 10 projects including the National Natural Science Foundation of China, the National Science and Technology Support Program, the National Key Research and Development Program, the Strategic Priority Research Programs of Chinese Academy of Sciences, as well as various local plans. He has published more than 50 papers in journals such as *Cities*, *Land Use Policy*, *Environment and Planning B*, *Computers, Environment and Urban Systems*, *Geographical Journal*, *Geographical Research*, and *Geographical Science* in both SCI and SSCI journals as well as Chinese core journals. He also has developed a couple of software and applied a few invention patents.

Chapter 1
Introduction

1.1 Concepts Related to Cities and Urban Space

1.1.1 The City and Its Origins

Literally, the term 'city' is a compound word of 'city' and 'market'. 'city' represents a defensive structure for guarding land with weapons, while 'market' refers to a place for trading. However, not all settlements with defensive walls are cities; some villages also have defensive walls (Wu and Li 2010). The emergence of cities is a symbol of human maturity and civilization, and a higher form of human communal living. Formally, there are two types of city origins: 'top-down' and 'bottom-up' (Luo 2007). In 'top-down' cities, military and political factors play a major role, with markets developing on the basis of cities. This type of city is often found in strategic and border locations, such as Tianjin Wei (Tianjin), Weihai Wei (Weihai, Shandong), Lingshan Wei (Lingshan Wei Town, Qingdao, Shandong), etc. In 'bottom-up' cities, economic factors dominate. These cities are more common and are the product of human economic development to a certain stage, essentially serving as human trade and aggregation centers. Regardless of the complexity of their formation process, the essence of cities boils down to these two types.

In general, a city can be defined as a concentrated settlement of non-agricultural population primarily engaged in industrial and commercial activities. It is also the center of social, economic, and cultural activities within a certain area, a complex and dynamically changing mega-system where various sectors and elements inside and outside the city are organically combined, characterized by a high density of population, buildings, and information (Wu and Li 2010).

In primitive societies, humans lived in caves and groups without forming fixed settlements. With human progress, humans created tools, improved their survival abilities, and began relying on fishing and hunting for food, forming primitive matrilineal communities. With the advancement of productive forces, humans began concentrated cultivation of edible plants, leading to agriculture; and the discovery of

F. Niu, *Urban Spatial Evolution Simulation*, Urban Sustainability,
https://doi.org/10.1007/978-981-97-3481-8_1

docile animals in hunting led to animal husbandry, marking the first major division of labor. Human life and the development of agriculture are inseparable from water, hence primitive settlements were mostly near water sources. To protect against wild animals and other tribes, humans-built walls and fences around their settlements with earth, stone, and wood, thus forming the rudiments of cities. When a tribe's productivity level gradually improved, life needs diversified, and labor division strengthened, specialized craftsmen (the second major division of labor) emerged, changing the characteristics of settlements: those primarily engaged in agricultural activities were rural areas, while those centered around commerce and handicrafts were cities. With surplus products came private ownership, the disintegration of primitive production relations, class differentiation, and the transition into slave society. Cities can be said to have emerged with private ownership and class differentiation during the transition from primitive to slave society. The emergence of cities in various ancient civilizations occurred at different times but all during this stage of social development (Wu and Li 2010).

Currently, there are three main academic theories about the origin of cities: first, the defense theory, which suggests that the purpose of building city walls was to defend against foreign enemies; second, the market theory, which posits that with the development of social production, people had surplus agricultural and livestock products, requiring a place for exchange. The places for exchange gradually became fixed, attracting more people, and early market forms evolved into cities; and third, the social division of labor theory, which asserts that with the continuous development of social productive forces, tribes developed specialized groups engaged in agriculture, handicraft, and commerce. Those engaged in handicraft and commerce needed a place to concentrate for trading goods, leading to the emergence and development of cities over time. Currently, scholars widely agree that 'cities are a product of the development of industry and commerce'.

1.1.2 Urbanization

Urbanization, in the context of industrialization and modernization, refers to the continual increase and expansion of urban population and size (Zhang 2008). Urbanization, also known as urbanization, involves the transformation of agricultural populations into non-agricultural populations, rural areas into urban areas, and agricultural activities into non-agricultural activities. Typically, the proportion of the urban population to the total regional population is used to reflect the level of urbanization. Additionally, urbanization can be divided into tangible and intangible forms. Tangible urbanization refers to the physical and formational aspects of urbanization, including population concentration, spatial form changes, and socio-economic structural transformation; while intangible urbanization pertains to the spiritual, consciousness, and lifestyle aspects of urbanization.

After the Industrial Revolution, as agricultural and pastoral people continuously flocked to new industrial centers, cities experienced unprecedented development, and the process of urbanization worldwide significantly accelerated. The first phase of urbanization in Europe began around the nineteenth century, including traditional Western European developed countries and most countries in Central Europe, and was essentially complete by the eve of World War I (Meng 2021). During this period, the European population grew from 20 million in 1800–150 million in 1900, reaching 344 million after World War II; the urbanization rate increased from 10% in the early nineteenth century to 50–67% by the mid-twentieth century. In other words, Europe essentially completed its urbanization process by the eve of World War II. Urbanization is not only a sign of affluence but also a symbol of civilization. The deep urbanization post-war is a development and deepening of this foundation. With the acceleration of urbanization, more and more cities have sprung up, and there are significant differences in their causes, functions, and types.

Currently, China is in a rapid phase of urbanization, characterized by several features: rapid urbanization speed, generally government-led with relevant policies, and developmental imbalances, lack of coordination with industrial development, and urban group development phenomena during the process. As the urban population size increases, as the physical carrier and result of urban residents' economic and social activities, urban space undergoes a series of changes and reorganizations, including spatial expansion and structural differentiation and polarization. In China's rapidly developing cities, urban space is experiencing expansion, differentiation, and other changes, impacting urban social development.

Data released by the National Bureau of Statistics on May 11, 2021, from the seventh census, show that the urban population was 901,911,62 people, accounting for 63.89% of the total national population. This result indicates that China has undergone a rapid process of urbanization over more than forty years of reform and opening up.

Undeniably, China's rapid urbanization has greatly promoted socio-economic development, but it also faces various issues such as continuous urban expansion, increasingly scarce land resources, and the need for enhanced ecological environmental protection. The key to addressing these issues lies in improving the quality of urbanization, ensuring that urban spatial structure evolution follows market economic laws, urban space utilization is scientific and orderly, industrial structure adjustments and layouts are rational, and urban spatial structure planning and layout are in line with urban functional requirements. How to make urban economic spatial structure more efficient, social spatial structure of higher quality, environmental spatial structure more harmonious, and organically integrate these aspects to guide the long-term sustainable development of cities are issues that need our research and exploration (Wang 2015).

1.1.3 Urban Space

For a long time, "space" has been a primary subject of study in geography. Since the 1990s, economists like Krugman have incorporated space into the analytical framework of economic models, leading to the emergence of "New Economic Geography" (He et al. 2014). This indicates that the concept and role of "space" are receiving increasing attention in the field of economics. Urban space refers to a spatial collection formed by a group of elements interacting in a specific geographical area, characterized by functional complementarity, regional characteristics, and structural orderliness. Through urban space as a medium, these elements continuously exchange and transmit material energy and information with their surroundings or within themselves.

Urban space is the projection of the organization of urban functions on geographical space, reflecting a combination of the city's political, economic, social, cultural life, natural conditions, engineering technology, and architectural space (Lu and Jiang 2020). Accompanying economic development and improvements in transportation conditions, they constantly change their structural forms and mutual positional relationships, with land use forms reflecting the evolution and characteristics of urban spatial structure. Urban space, unlike rural space, contains more elements, and the connections between spatial elements are closer. These social, economic, political, and cultural elements collectively support the normal and coordinated operation of the city. The definition of urban space varies with different research perspectives and focuses. For example, from a geographical space perspective, it mainly focuses on the physical attributes of the city, i.e., the projection of various material factors on urban land. The social space perspective emphasizes the social attributes of urban space, like social classes, neighborhoods, and community organizations. The economic space perspective focuses on the scope and content of urban economic activities, usually extending beyond the city's geographical space. The psychological space perspective understands and studies urban spatial attributes from the perspective of human cognition and perception (Wang 2007). Therefore, urban space has connotations of perceptual and imaginal space, and the formation of perceptual space is determined by social-cultural characteristics and influenced by social processes (Huang 2002). Additionally, urban space includes mathematical space, ecological space, and more.

In the field of urban planning, the main subjects of urban space research are urban spatial structure, urban spatial form, and urban spatial development laws. Urban planning in China originated from architecture, so the architectural concept of space has deeply influenced the planning community. This concept emphasizes the physical attributes of space, neglecting its social attributes. Even when addressing the social aspects of space, it tends to view space's "sense of place" from an environmental behavior perspective. Grasping the complexity of space's essence requires placing it within a complex social system, rather than abstracting it simplistically. The social attributes of space determine its close relationship with socio-economic elements. Space is not an entity distant from politics and ideology; it has always been changed and reconstructed by history and various natural forces, a process inherently political. Therefore, urban space is not just the material space in the traditional sense but emphasizes social processes, cultural experiences, and behavioral meanings of space (He 2006). Modern urban spatial planning is an application of planning and design after studying social needs, economic development, cultural traditions, behavioral laws, visual psychology, and policies and laws (Duan 2005).

1.1.4 Urban Space Characteristics

Urban space is the carrier of urban life, providing suitable places for various resident activities. Urban space has unique properties that allow different social lives and cultural traditions to communicate and coexist. The characteristics of urban space generally include publicness, placeness, bearing capacity, complexity, and evolutionary nature (Kang 2012).

Publicness. Urban space is a public place that provides production and living services, closely linked to the lives of citizens. "Publicness" also implies shared interests and ownership, meaning urban space belongs to no private group or individual but is shared by every individual in the city.

Placeness. The placeness of urban space undergoes a process of gradual recognition, encompassing time, space, and behavior. Urban space accommodates the material needs of city life, bearing the social needs of various urban activities. Residents use various facilities to participate in urban activities, developing an identification with urban spaces over time. Different urban spaces have different urban functions, and the recognition of these functions comes from people's long-term participation in urban activities provided by these functions.

Bearing Capacity. Urban space is a carrier. Its bearing capacity is manifested in its physical, social, economic, and ecological aspects. Urban space enables the material existence of the urban system and coordinates various materials in space; it provides places for socio-economic activities of city residents; it is a carrier of various economic elements in the city; urban space is an organism, developing based on ecological environments (Zhang 2003).

Complexity. Urban space is "complex," with its complexity evident in both spatial and temporal dimensions. In terms of spatial dimensions, a city is a collection of multiple functions, such as living, working, transportation, consumption, etc. A single urban space unit usually has multiple functions, and the functions it carries change and succeed over different periods.

Evolutionary Nature. From the long history of the origins and development of cities, we find that under the interaction of complex internal and external contradictions, cities continuously develop, and urban space evolves from primitive simple community space to complex urban space. The comprehensive effects of various factors bring about the imbalance of urban space. Imbalance generates dynamics, causing continuous operation of urban space; meanwhile, as urban functions become increasingly complex, the urban system evolves towards more advanced forms, guiding the development of urban space towards more advanced forms (Wang 2000).

1.1.5 Urban Spatial Structure

Urban spatial structure refers to the geographical location and distribution combination of various functional areas in a city, the spatial and geographical projection of urban functional organization (Shi 2004). Urban spatial structure is a complex and diverse concept that includes not only the linear structure formed by urban physical facilities but also relatively stable structural contents such as urban economic structure, social structure, and environmental structure; it is both an objectively existing concept and a subjective consciousness (Hu 1995). It is influenced by economic, social, and environmental elements (Fu 2012). From an expression perspective, urban spatial structure is the tangible manifestation of the plane and three-dimensional forms, styles, and layouts of various material elements in a city, a spatial combination pattern of various architectural forms. In essence, it is a material form of complex human economic, social, and cultural activities in the historical development process, a comprehensive reflection of the interaction between human activities and natural factors in a specific geographical environment, a specific representation of urban functional organization methods in space. The configuration of urban spatial structure significantly affects the overall spatial organization efficiency.

Research on urban spatial structure can be divided into three levels: micro (main urban area), meso (urban and suburban areas), and macro (urban group space). The internal and external spaces of the city are the main parts of the evolution of urban space. The internal spatial structure of a city is the functional zoning structure of land within the built-up area of a city (usually referring to the urban area), a regional structure formed by the internal differences in land use connected by various activities. The external space of a city includes suburban satellite cities, specialized towns, and their surrounding rural hinterlands, areas where internal urban space grows and expands outward. In terms of expression, urban spatial structure mainly includes the density and location (layout) of various material entities and urban morphology (Guo and Jiang 2002).

1.2 Changing Urban Space in China

1.2.1 Impacts of Urbanization

Urban space continually evolves over time. In the context of rapid urbanization, changes in urban space are quicker and exhibit many new characteristics. Urbanization has led to informatization, globalization, and networkization, forming new socio-economic development models. It has affected regional economic patterns, fundamentally altering urban and regional spatial development models and influencing the future development of cities.

Urbanization has expanded urban spatial scales, leading to population growth, expansion of urban land, intensification of socio-economic activities, and closer external interactions, directly changing the scale of urban development. However, disorderly and blind expansion can cause various urban problems, even crippling cities.

Urbanization changes the original urban spatial functions. Firstly, it directly leads to changes in land use. The increase in urban population alters the original living patterns, necessitating more residential space for the growing population. This will expand urban land use, changing both the nature of land use and existing residential forms. Secondly, as large numbers of rural populations migrate to cities, they provide labor for urban development, accelerating urban industrial restructuring and affecting urban spatial structure. Additionally, urban development is inevitably constrained by various factors, leading to historical planning issues that require readjustment in subsequent development.

1.2.2 Evolution of Urban Space in China

With the continuous increase in China's urbanization rate, the evolution of urban space has also accelerated. This evolution is first evident in the number of cities and the expansion of urban spaces. From 1978 to 2020, the number of cities in China rapidly increased from 193 to 684, and the number of megacities (with a population of over 5 million) rose from 13 to 91. Meanwhile, the urban area also expanded significantly. For example, the urban district area grew from 178,692 km^2 in 2010 to 200,570 km^2 in 2019; the built-up area expanded from 40,058 km^2 in 2010 to 60,312 km^2 in 2019; and urban construction land area increased from 39,758 km^2 in 2010 to 56,076 km^2 in 2018. From 2000 to 2019, the proportion of urban population in China's total population (urbanization rate) increased from 36.2 to 60.60%, and the rural population's share decreased from 63.78 to 39.40%. The steady growth in urbanization reflects the gradual migration of rural populations to cities. The economic development also validates the acceleration of urbanization, with urban economies primarily based on secondary and tertiary industries, while rural economies are dominated by primary industries like agriculture, forestry, fishing, and animal husbandry.

Analysis of nighttime light data from 2001 to 2013 suggests that urban spatial expansion in China is mainly concentrated east of the Hu Huanyong Line in economically developed areas and is becoming more contiguous, especially in regions like the Pearl River Delta, Yangtze River Delta, and Beijing-Tianjin-Hebei. In contrast, the expansion west of the Hu Huanyong Line is relatively smaller and more fragmented, with some cities experiencing rapid expansion, like Ordos in Inner Mongolia, where the built-up area expanded nearly fivefold during the study period (Wang 2020).

During development, urban space evolves with differentiation and polarization. According to the theory of unified social space, social production and space production interact and influence each other; thus, socio-economic changes inevitably affect the transformation of urban spatial structures. Accompanying the commodification of housing, residential differentiation and segregation have led to spatial clustering and differentiation of social classes, gradually altering China's traditional urban residential patterns (Zhang 2008). The impact of the unit system on urban space is diminishing, and research in major Chinese cities like Shanghai, Beijing, and Nanjing indicates that Chinese cities are experiencing spatial differentiation (Wu 2001). Due to the restructuring and transition of urban industrial structures and the stratification of the urban labor force, income disparities have widened, leading to spatial polarization patterns between rich and poor in urban spaces. This kind of polarization typically occurs in "global cities" like New York, London, and Tokyo. In China's megacities, such as Beijing, Shanghai, and Guangzhou, under the backdrop of globalization, new international division of labor, and global economic restructuring, spatial fragmentation and "dual city" trends are emerging, with increasing social spatial segregation and opposition (Yang 2006).

Despite the continuous expansion of urban areas in the process of urbanization, cities remain attractive due to better infrastructure and resource allocation, leading to high population density and crowded spaces. China faces challenges with high population density, especially in the eastern regions, defined by the three-step geographical features from east to west. According to the seventh population census data, the population density in Tianjin's Heping District has reached 35,000 people per km^2, and the central urban areas of Beijing and Shanghai have densities of 24,000 and 23,000 people per square kilometer, respectively, much higher than the population density in cities perceived as densely populated, like Seoul (16,000 people per km^2) and Tokyo (14,000 people per km^2). The uneven development of cities has led to inadequate planning, poor living environments, and extremely low per capita living areas in lower social strata's residential spaces, particularly in "urban villages," shantytowns, and areas densely populated by migrant workers. These areas, often strictly controlled by policies, cannot be rebuilt but only expanded on their existing basis. Hindered vertical development leads to horizontal expansion in streets and alleys, resulting in the occupation of public spaces, reduced distance between houses, safety hazards, and loss of the spatial texture and aesthetic that cities should pursue.

1.2.3 Basic Forms of Urban Space Evolution

Driven by internal and external developmental forces, the expansion and evolution of urban space essentially represent the changing distribution pattern of urban activities. This evolution includes not only the expansion of urban land use in horizontal directions and extensions vertically into the air or underground, but also the addition of urban elements, and the composition and transformation of urban structural functions. The evolution of urban space can primarily be categorized into the following types: simple evolution, structural evolution, and distribution evolution.

Simple Evolution: In the initial stages of development, the growth of urban space mainly focuses on expansion in scale, which is reflected by two main indicators: population growth and land use expansion.

Structural Evolution: The development of a city is a gradual process from simplicity to complexity. As the evolutionary process progresses, the developmental elements of the city change, and these changes in element structure reflect the transformation of urban space.

Distribution Evolution: As the internal development and refinement of a city continue, the interactions and interconnections between cities strengthen, reflecting changes in the external system of urban space and the various relationships between cities (Gu et al. 2000).

These forms of urban space evolution highlight the dynamic and multi-dimensional nature of urban growth and development, shaped by a combination of demographic, economic, social, and physical factors.

1.2.4 Dynamics of Urban Space Evolution

The dynamics driving the evolution of urban space include both internal and external forces.

1. Internal Dynamics of Urban Space Evolution

Internal dynamics generally stem from the openness, non-equilibrium, and mutability of urban space development (Kang 2012).

Openness of Urban Space Development: Cities are typical open systems, continuously interacting with the external world (material flow, information flow, capital flow, etc.). As the factors influencing the system increase, urban space continually evolves and develops. The openness of the urban space system carries the risk of leading to unordered development, but this risk is mitigated by defining the boundaries of the city and filtering external inputs (Zhang 2003).

Non-equilibrium of Urban Space Development: In the process of exchanging with the external environment, the complex external environment keeps the elements of urban space development in a constant state of imbalance. This imbalance is the source of the dynamics of development. The unbalanced effects on urban space keep it in a state of continuous developmental motion. Though always moving towards a state of balanced forces, the state of force in urban spaces continually changes due to the intervention of external factors and internal force changes, keeping urban space dynamically developing. Dissipative structure theory posits that non-equilibrium is the source of order; all balanced systems will lose their capacity to maintain development and become static "dead" systems.

Mutability of Urban Space Evolution: Due to the uncertainty of external and internal forces, we can predict the main trajectory of urban space evolution, but not the specific state of development at a particular stage. Minor changes in a variable might cause a significant shift in urban space development, similar to the "butterfly effect."

2. External Dynamics of Urban Space Evolution

The external dynamics mainly stem from two aspects: the requirements of economic development and environmental constraints. These are the primary external drivers of urban space construction and development and reflect the main manifestations of social needs.

Requirements of Economic Development: The development of a city's economy determines the pace of urban space evolution. Economic growth guides urban development, with the immense economic benefits from urban land development leading urban space to evolve in directions that create more benefits. This also accelerates urbanization, drawing more surplus rural labor into cities, and thereby converting rural populations into urban residents. With cities expanding like "spreading pancakes," some rural lands are transformed into construction lands, forming a simple evolution of urban space, marked by an increase in urban population and land use expansion. Economic development also leads to adjustments in industrial structures, causing a reallocation of resources, including land.

Resource and Environmental Constraints: The development of urban space is constrained by resources and the environment. Each city has its carrying capacity for resources and the environment, and urban operation must occur within this capacity. The capacity to some extent dictates the direction and space of urban development. Improvement in the urban environment is beneficial for attracting capital and talent, thereby promoting local economic development, improving the quality of life of local residents, and directly affecting urban economic development and the evolution of urban space. The development of a city also depends on the resources it possesses, with the city's industrial structure and spatial structure adjusting around these resources.

1.3 Chapter Summary

Cities are not only the products of human civilization but also continuously shape it. Having evolved over thousands of years, their functions and forms have diversified. Alongside the development of cities, human understanding and research on them have deepened. Serving as an introduction to the book, this chapter reviews the origins and developmental history of cities. It elaborates on concepts such as urbanization and urban space, and then briefly analyzes the impacts of urbanization, the patterns of urban spatial evolution in China, and their driving mechanisms. The goal is for readers to gain a macro understanding of cities and their evolutionary processes, laying the groundwork for subsequent chapters that further discuss the evolution of urban space and the simulation of urban spatial evolution. For those interested in delving deeper into urban topics, additional reading materials in urban geography are recommended.

References

Duan J (2005) Thinking on the problems of China's urban planning theories and practices. Urban Plan Forum 01:24–27 (in Chinese)

Fu L (2012) Transformation of metropolitan spatial structure and its evolution. China Architecture & Building Press, Beijing, pp 70–74 (in Chinese)

Gu C, Zhen F, Zhang J (2000) Agglomeration and dispersion. Southeast University Press, Nanjing, p 3 (in Chinese)

Guo H, Jiang M (2002) Urban spatial economics, vol 6. Economic Science Press, Beijing, pp 21–23 (in Chinese)

He Z (2006) Theories and research progress in urban spatial planning in China. Planner (07):87–90 (in Chinese)

He C, Guo Q, Ma Y et al (2014) Progress of economic geography in the West: A literature review. Geogr Res 69(08):1207–1223 (in Chinese)

Hu J (1995) Chinese cities: patterns and evolution. China Architecture & Building Press, Beijing, pp 78–80 (in Chinese)

Huang Y (2002) Urban space theory and spatial analysis. Southeast University Press, Nanjing (in Chinese)

Kang H (2012) The reciprocity between urban infrastructure and space evolution. Harbin Institute of Technology, Harbin (in Chinese)

Lu D, Jiang M (2020) The evolution and governance strategies of urban space in the era of the internet. Nankai J (philos Literat Soc Sci Edn) 02:92–100 (in Chinese)

Luo L (2007) The type and origin of Chinese ancient cities. J Yanbian Univ (soc Sci Edn) 02:87–91 (in Chinese)

Meng Y (2021) Analysis of the urbanization process in Europe, 1870–1940. Urban Culture Stud (02):31–45 (in Chinese)

Shi S (2004) The dynamics of urban spatial structure's evolution. Urban Plan Rev (01):50–52+96 (in Chinese)

Wang P (2000) Systematic study of urban public spaces in China. Tsinghua University, Beijing, p 11 (in Chinese)

Wang C (2007) Study on the interaction mechanism of urban transport and urban spatial evolution. Beijing Jiaotong University, Beijing (in Chinese)

Wang J (2015) Research on Shanghai urban spatial structure evolution. Jilin University, Jilin (in Chinese)

Wang S (2020) The analysis on the spatiotemporal characteristics and driving factors of urban spatial expansion in China. Zhongnan University of Economics and Law, Wuhan (in Chinese)

Wu Q (2001) Theoretical and practical study of residential space differentiation in large cities. Science Press, Beijing (in Chinese)

Wu Z, Li D (2010) Principles of urban planning, 4th edn. China Architecture & Building Press, Beijing (in Chinese)
Yang S (2006) Evolution of social space in large Chinese cities. East China University of Science and Technology Press, Shanghai (in Chinese)
Zhang Y (2003) Research on self-organization of urban space development: a case study of Shenzhen. Southeast University, Nanjing (in Chinese)
Zhang X (2008) The brief analysis of Chinese city space evolution during urbanization. Soc Work Bi-Mon (Theory) 05:48–50 (in Chinese)

Chapter 2
Overview of Urban Spatial Evolution

The study of urban space and its evolutionary processes is a classic subject in urban geography, with a long history of academic accumulation. In this chapter, by reviewing the research achievements of predecessors, we elucidate the relevant concepts and theories of urban space, as well as interpret the main influencing factors and mechanisms of urban spatial evolution. This sets the groundwork for subsequent chapters to further clarify the methods of simulating urban spatial evolution.

2.1 Basic Concepts of Urban Spatial Evolution

Urban space, varying in connotation across different research fields, can generally be divided into micro and macro levels. Micro-space refers to the space occupied by buildings and their enclosed spaces, while macro-space encompasses the broader spatial domain of a city, including three-dimensional space (Sun 1997). From a macro perspective, urban space is essentially a collection of spatial elements with specific regional characteristics and functional structures. It's a dynamic, hierarchical, adaptive open system primarily interconnected by flows (information, materials, energy, etc.) (Tang 2016). In cities, dynamic interactions occur among different species and communities. Through complex self-organizing activities, they collectively form a variety of urban relationships, which in turn influence the overall morphology of urban space (Huang 2002), manifesting in land use and architectural environmental differentiation.

The concept of urban spatial evolution has been enriched with urban development. In a narrow sense, it mainly refers to the expansion and structural evolution of urban construction land (referred to as "urban land") (Zhu and Yao 1999). Urban land scale denotes the size of the land occupied and used by a city, encompassing not only the increase in the amount of urban land but also changes in expansion patterns, directions, and speeds. It represents a "quantitative change" in the urban

F. Niu, *Urban Spatial Evolution Simulation*, Urban Sustainability,
https://doi.org/10.1007/978-981-97-3481-8_2

development process. Land structure directly reflects the urban land use status within a certain period and profoundly influences the urban economy, society, and functional construction. Urban spatial evolution, in this sense, is specifically manifested in the expansion of urban land scale, changes in the quantity distribution ratio, and spatial structure. It occurs through functional replacement in old urban areas (intrinsic expansion) and by developing new industrial areas, development zones, and residential areas (extrinsic expansion) (Zhu and Yao 1999).

In a broader sense, urban spatial evolution reflects not only land aspects but also changes in urban population size, industrial structure, etc. For example, with ongoing urbanization, the transformation of non-agricultural to urban population accelerates, leading to changes in labor employment structures; urban functional zones are more in line with modern urban development needs, shifting from single-center distribution to multi-center; urban economic industry structure softens, transitioning from manufacturing to service industries, bringing about a reorganization and adjustment of urban industries; urban diffusion and agglomeration coexist, forming a networked urban system. On one hand, urban spaces continuously expand outward, while on the other, central urban areas strengthen their agglomeration functions, and the connections between satellite cities and surrounding cities become increasingly close, making urban evolution more complex. The emergence of concepts like "ecological spatial structure" and "sustainable development cities" means that cities not only serve as carriers for economic development but also provide an elegant environment for residents' production and living (Hou and Xi 2004).

Overall, urban spatial evolution reflects both the spatial distribution patterns of material elements and activities, and their interactions. Superficially, it is the tangible manifestation of various material elements in the city in terms of plane and three-dimensional forms, styles, and layouts – a spatial combination of various architectural forms. In essence, it represents the material form of complex human economic, social, and cultural activities in historical development, a comprehensive reflection of human activities and natural factors under specific geographical conditions, and a specific representation of urban functional organization in space (Hu 1995). Besides the linear structure composed of urban material facilities, it also includes relatively stable structures such as urban economic structure, social structure, and environmental structure. Thus, urban spatial structure is both an objective concept and a manifestation of subjective consciousness (Fu 2012). Urban economic and population development lay the foundation for urban spatial evolution. As the city increasingly serves as a carrier for economic development, planning for the expansion of urban construction land and spatial structure should consider the needs of urban life and production and related natural factors, while also considering the "future development space of the city".

As an important basis for human understanding of urban development laws and urban planning, the study of urban spatial evolution has always been a hot topic in urban geography. From a research scale perspective, it can be divided into micro (inner urban space), meso (outer suburban space), and macro (regional urban group space) levels, with inner and outer urban spaces being the main bodies of urban spatial evolution. The inner urban space is a regional structure formed by the internal differentiation of urban internal functions and various activities, also known as the city space; the outer urban space includes suburban satellite cities, specialized towns, and their surrounding rural hinterlands, where the growth and outward expansion of the inner urban space occur (Guo and Jiang 2002). From the perspective of urban space attributes, research on urban spatial evolution involves both the physical attributes of urban space, i.e., the spatial differentiation and evolution process of the urban physical environment, and the social attributes, i.e., the spatial distribution and evolution process of urban society, economy, etc. (Tang 1997). The physical attribute of urban spatial evolution is mainly manifested in the expansion of urban scale and changes in urban spatial organization patterns, while the social attribute of spatial evolution mainly focuses on three aspects: complex changes and recombination of urban internal functional spaces (Zhang 2001); the trend of "economic service" in urban economy, i.e., the so-called "softening" of urban industrial structure; and the outward movement of urban residents' living spaces and the "suburbanization" of important industries. In fact, both urban physical space evolution and urban social space evolution are reflected in the scale and functional changes of urban land use, i.e., in urban planning (Hou and Zhang 2006).

China's urbanization process is in a period of accelerated development, with cities in a phase of rapid growth, characterized by the continuous expansion of urban population size and land use, and constant adjustments in urban industrial and land use structures. Urban spatial evolution not only meets the needs of industrial upgrading and transformation for urban space but also promotes functional replacement in cities, maintaining urban development vitality. Meanwhile, the expansion of urban land scale and changes in land use structure have led to the reorganization of urban regional structures, causing profound changes in urban economic, social, and functional structures, decisively influencing the urbanization process. Whether it promotes or delays urbanization depends on whether it follows the objective laws of urbanization and urban spatial evolution and whether the urbanization process is in harmony with the urban spatial evolution process. Therefore, the issue of urban spatial evolution is a very important one in China's urbanization process (Zhu et al. 2000).

2.2 Relevant Theories of Urban Spatial Evolution

2.2.1 Ebenezer Howard's Garden City Theory

The Garden City concept was proposed by British scholar Ebenezer Howard in his 1902 work "Garden Cities of Tomorrow." Howard envisioned the Garden City as a fusion of urban and rural elements. He believed that the scale of the city must be limited, allowing every household to have easy access to rural natural spaces (Butlin and Howard 1899). This idea played a significant enlightening role in modern urban planning. Howard made epoch-making insights into urban planning issues such as urban density, spatial layout, and green vegetation, greatly influencing subsequent theories like "Organic Dispersion Theory" and "Satellite Town Theory."

Furthermore, Howard's Garden City concept first introduced a new model of integrated urban–rural development. It suggested that a Garden City should combine the advantageous aspects of both urban and rural areas without the disadvantages of either. He argued that only by integrating urban and rural areas could complex and severe social and environmental issues be comprehensively resolved. This concept closely aligns with the principles of integrated urban–rural planning proposed in China. In the context of the social background of that time, the Garden City idea was a revolutionary change in Western societal life, making Howard one of the most influential urban planning theorists in modern Western history.

2.2.2 *Le Corbusier's Modern Urban Theory*

Le Corbusier was the most famous architectural and urban planning master in France in the twentieth century. In 1922, he published the "City of Tomorrow" plan: The core idea of the theory was to change the entire social, cultural, and geographical environment through systematic transformation of the physical space. In addressing urban congestion, Le Corbusier sought to reorganize the structure of large cities. Based on further concentration of population, he proposed rational urban functional zoning, suggesting the construction of numerous high-rise buildings, subways, and efficient transportation systems with segregated pedestrian and vehicle traffic within the city; he did not advocate the rebuilding of satellite towns in the suburbs. In fact, this concept improved some of the problems encountered in urban development to a certain extent. However, this idea of concentrated development through steel and concrete high-rise buildings also destroyed the cultural and historical landscape of cities, making them monotonous and lacking in human warmth.

2.2.3 *Burgess' Concentric Zone Theory*

In 1923, American sociologist Ernest Burgess proposed the concentric zone model. Burgess primarily analyzed urban areas based on the types of residential populations and districts. He believed that five forces—centripetal, specialization, separation, centrifugal, and centripetal-centrifugal—caused geographic differentiation in cities. In this process, various urban zones continuously encroached upon and shifted, forming a concentric circle pattern of expansion. The main flaws of this model are: first, it did not consider the intersection between different areas and the role of urban transportation; second, it failed to take into account the layout of industrial activities, which are a major vitality of cities, and their impact on urban land use. Burgess noted that the city center is the Central Business District (CBD). The CBD, as the center of modern cities, has become a widespread norm and, in turn, influences urban planning and renewal, especially offering a corrective direction to the government-centered urban layout in traditional Eastern cities. In China, some cities are planning to relocate administrative institutions from the city center to develop business functions. In this sense, Burgess's model still has some relevance for monocentric cities.

2.2.4 *Hoyt's Sector Model*

Hoyt conducted a study on urban rent in American cities, examining nine tendencies in the layout of urban residences and concluding that urban land use follows a sector model. Hoyt's model highlights three issues: first, the city center is characterized by the Central Business District (CBD); second, lower-class residential areas intersect and mix with wholesale and light industrial zones; and third, different grades of residential areas are distributed by sectors rather than by distance from the center. In his model, the functional zoning of urban layout is already emerging, and urban transportation, especially the traffic lines of the central CBD and external transportation lines, greatly influence the urban layout.

2.2.5 *Harris and Ullman's Multiple Nuclei Theory*

In the 1940s, Quin proposed that the Central Business District (CBD) is the main center of a city, but there are also other centers, each influencing a certain geographic area. Building on this, Harris and Ullman studied the regional structure of various cities, arguing that the diversification of urban centers and the differentiation of urban spatial structure are formed by four processes: (1) the locational process of various industries aimed at their own interests; (2) the process of generating agglomeration benefits; (3) the separation process arising from the contrasting interests of different industries; (4) the influence of land prices and rents on industrial location. On this basis, Harris and Ullman proposed the multiple nuclei model. In their model, the CBD remains the center of urban activities but is not necessarily the geometric center, often leaning to one side. Wholesale areas and low-pollution light industries oriented towards the local market are located near the CBD. Lower-class residential areas still cluster around the CBD and wholesale areas, while middle and upper-class residential areas are clustered on the other side, interspersed with numerous public cultural facilities. Industries are mainly distributed in the outer suburban areas, near separate residential areas. In the multiple nuclei model, urban land prices do not simply decrease from the center to the periphery; instead, several peak areas emerge. In the early stages of backward cities, land prices continuously decrease from the center to the edge. However, in the era of multiple nuclei, apart from the CBD, cities have other sub-centers, making the multiple nuclei model more characteristic of modern cities.

2.2.6 Christaller's Central Place Theory

Central Place Theory was first introduced by German economic geographer Walter Christaller in his 1933 book "The Central Places in Southern Germany". Christaller believed that the formation of an urban hierarchy is governed by three principles: the marketing principle, the transportation principle, and the administrative principle. Under the influence of these principles, central place spatial networks exhibit different structural characteristics, and there is a strict hierarchical order of central places and the size of their service areas, which can be arranged into a regular and tight series.

Christaller initially proposed the market as a spontaneously formed rural market, a central place developed and clustered according to capitalist free competition relations. The structure and layout of lower-level central places, generally dominated by a market economy, are appropriately explained by the marketing principle. The transportation principle is more suitable for newly developed areas where transportation networks are distributed linearly. It is more likely to appear in reality and has a greater impact than the marketing principle. Higher-level central places, which require long-distance transportation, are generally arranged according to the transportation principle. The administrative principle applies to areas with strong governing institutions or those based on administrative organization. In regions that are self-sufficient and relatively remote, the administrative principle plays a stronger role. Middle-level central places are more influenced by the administrative principle.

The variable hierarchy in regional models formed under various conditions cannot be described in fixed terms. Additionally, Christaller believed that consumers would first choose the central place closest to them. However, in reality, consumer behavior is governed by various needs, and consumers tend to engage in economic or social activities in higher-level central places. This leads to the continuous expansion of the market area of higher-level central places, distorting the structure of the central place system. However, these flaws do not diminish the great significance and historical status of the Central Place Theory. It is one of the theoretical foundations of regional economic studies. In studying regional structure, the relationship between central places and their hinterlands described by Christaller provides important guidance for the rational layout of regional public services and economic and social functions. Central Place Theory and the concept of antigravity attraction systems are used to explain the urban hierarchy within a region or country, its spatial distribution patterns and principles. Furthermore, some geographers and urban planning scholars believe that the service range of urban centers shares many similarities with Central Place Theory and the antigravity attraction system, and these concepts can be used to explain certain aspects of the scale and functional positioning of urban centers, as well as the distribution characteristics of urban commercial center systems (Christaller 1966).

2.2.7 Theoretical Concepts of the Machu Picchu Charter

In 1977, a group of urban planners and designers, starting from the principles of the "Athens Charter," signed a new declaration at Machu Picchu. This act demonstrated the vitality and continuity of the modern movement in architecture and planning. The "Machu Picchu Charter" indicated a shift in urban planning theory from the "functional segregation" advocated by the Athens Charter to "functional integration." It pointed out that functional segregation disrupted the organic connections of the city and overlooked the close relationships among urban residents.

The Machu Picchu Charter explicitly proposed that a city should not be viewed as a mere assembly of components, but rather, efforts should be made to create an integrated, multifunctional, and organic environment. This approach would enable the city to grow and develop within this dynamic, organic setting. The charter emphasized the dynamism and process-oriented nature of urban planning. The ultimate driving force behind urban development should be the patterns of human culture, social interaction, and political structures. It promoted the fundamental ideas of social culture and rejected the mechanistic material spatial determinism. The charter suggested a more holistic and human-centered approach to urban planning, focusing on creating spaces that foster community, cultural richness, and social interaction, in contrast to the more segregated and functionally divided spaces of earlier planning models.

2.2.8 Theories of Satellite Town and New Town Construction

The concept of satellite towns can be traced back to Ebenezer Howard's idea of the "Garden City." Satellite towns are extensions of the functions of the central city, surrounding the urban core, closely related to it, yet possessing a degree of independence. The first generation of satellite towns, also known as dormitory towns, served primarily as residential areas where residents' work and cultural life were still centered in the main city.

The second generation of satellite towns expanded in size. To facilitate residents' lives, the government equipped these towns with a certain number of public facilities. Some factories also relocated to these satellite towns, allowing some residents to work closer to home.

The third generation of satellite towns is essentially independent of the central city area, with more comprehensive urban functions. They not only offer significantly more employment opportunities but also provide greater convenience for daily life. These towns feature a full range of functions, including culture, education, leisure, entertainment, finance, dining, and more. With complete and modern central infrastructure, they are also referred to as "new towns." They can meet the work and life needs of residents on-site, functioning as independent cities with well-rounded capabilities, thus reducing dependency on the central urban area.

The evolution of satellite towns has progressed from the first generation's dormitory towns to the third generation's satellite new towns. The town population and geographical space have expanded from small to large, with the urban functions transitioning from single-purpose to comprehensive. The objective has also shifted from initially dispersing the population to reorganizing the urban layout and promoting long-term balanced development of the region.

2.2.9 Organic Dispersion Theory

The theory of organic dispersion was first mentioned by Finnish scholar Eliel Saarinen in 1942. The organic dispersion theory advocates that administrative and public institutions should remain in the city center, while industries should be moved out of the central area. It suggests that areas for daily activities should be concentrated, while locations for occasional activities should be dispersed. The theory emphasizes walking as the primary mode of transportation for daily activities, aiming to reduce traffic pressure and congestion.

The starting point of this approach is the reorganization of urban functions, implementing organic dispersion of the city to delay the decline of central areas. It calls for scientifically planned new town functions. Only through such measures can a city maintain its vitality and achieve healthy, orderly development. This theory reflects a holistic approach to urban planning, focusing on the balance between centralization and decentralization, integration of urban functions, and enhancing the quality of life for urban residents.

2.3 Factors Influencing Urban Spatial Evolution

The urban spatial evolution process is the transition of various urban activity spaces. This process includes changes in the scale of various urban activities and their spatial distribution. When the number of urban activities increases, changes in their spatial distribution patterns are accompanied by spillover, namely, the expansion of the city. The sites of urban activities are enterprises and households, therefore, urban activities can be roughly divided into these two categories. The classification of urban activities will be elaborated in subsequent chapters. The influencing factors of each type of activity differ, and so do their location choices. This book will discuss these in detail in later chapters when discussing the establishment of an "Urban Land Use-Transport Interaction Model." At the macro level, the factors influencing urban evolution can be summarized in several aspects.

2.3.1 Economic Factors

Different levels of economic development have varying impacts on urban spatial structure, making it one of the most active factors in creating contradictions between urban functions and spatial structures. Economic development not only disrupts the balance of the urban–rural spatial structure system but also prompts changes in external influencing factors such as technology, politics, and residence, thereby enhancing the adaptability of spatial structures. When the expansion of urban functions does not exceed the bearing capacity of the urban spatial structure, it remains in a stable state. However, when urban economic development leads to an expansion of urban functions beyond the bearing capacity of the spatial structure, adjustments occur in the urban structure, either partially or wholly. Consequently, new peripheral areas emerge, existing peripheries either develop or decline, and the spatial structure continuously adapts to changes in the economic system.

The specific impacts of economic development level on urban spatial structure are mainly reflected in the following aspects: First, improvement in urban construction funding conditions enables cities to increase investment in infrastructure, thereby enhancing the area's attractiveness to various urban activities. Improved funding also provides economic security for urban renewal and new district construction, leading to changes in the urban spatial structure. Second, the enhancement of urban economic levels promotes the upgrading of urban industrial structures, leading to new types of demand for urban land use, especially with the increasing proportion of the tertiary sector, dominated by information and service industries. This shift from a production-oriented to a service-oriented function requires a reorganization of urban spatial structure. Additionally, at the micro-level, the form of economic organization also influences changes in urban spatial structure. Under less socially advanced production organizational forms, with low internal division of labor and weak systemic organization without hierarchical and functional differentiation, this economic organization leads to an urban spatial structure with unclear functional zoning and a lack of spatial hierarchy. In contrast, in more socially advanced economic organizational forms, with clear social division of labor and different hierarchical systems, this is reflected in urban spatial structures through the emergence of functional zoning and multi-core urban spatial structures.

Furthermore, the cyclical nature of economic development speed determines the periodicity of urban spatial expansion (Yang and Zhang 1997). Urban spatial expansion is not a gradual and uniform outward progression but exhibits cyclical changes that vary with economic fluctuations. During phases of rapid economic development, urban space expands quickly, whereas economic slowdowns lead to stagnation in urban spatial expansion. Practice shows that during periods of rapid economic growth, urban spatial expansion mainly manifests as outward extension beyond the built-up area, with urban land in a dispersed state and a decrease in the compactness index (the ratio of land scale to the area of the city's smallest circumscribing circle). During stable or slow economic growth periods, urban development shifts from extensional to intensional, focusing on internal infilling and redevelopment, resulting in an increase in the urban agglomeration compactness index.

2.3.2 Demographic Factors

The spatial movement of urban populations is an extremely important aspect of the evolution of urban spatial structure (Feng and Zhou 2002). The internal migration of urban populations has specific directions and distances, and long-term migration alters the population numbers in both the areas of influx and outflow, thereby affecting the overall pattern of population spatial distribution in the city. Additionally, migrants differ in age, occupation, income, and educational level. Long-term migration results in the aggregation of people with similar social characteristics in certain areas, leading to the formation and maintenance of urban communities. As populations move and the spatial distribution of populations changes, the formation of communities and the corresponding changes in regional functions also occur.

2.3.3 Sociocultural Factors

Essentially, urban spatial structure is a complex physical manifestation of human social, economic, and cultural activities in space. Real urban spatial structures invariably bear the imprints of history, social systems, religious beliefs, social structures, and cultural types. Their evolution showcases distinctive characteristics shaped by historical, social structural, and cultural diversity. Therefore, social and cultural activities in cities have a significant impact on the formation and development of urban spaces. It can be said that urban space is a carrier of regional culture, and every urban space is formed under the long-term accumulation and influence of culture. In the dynamic process of changes in urban spatial structure, it is human culture that shapes the spatial structure of cities. In turn, the evolving urban spatial structure influences human behavior (Tao and Chen 2001; Zheng and Lin 2002).

2.3.4 Urban Intelligence and Scale Factors

Urban function refers to the status and role of a city in politics, economy, and culture within a certain region. It directly influences the city's size, land use composition, industrial structure, and these in turn have a significant impact on urban spatial structure. Additionally, the size of a city determines the radius of urban activities and modes of transportation, which are decisive factors for the urban spatial structure. Therefore, cities with different functions and sizes generally have different spatial structures. Under certain size conditions, concentric circle layer expansion can achieve higher agglomeration benefits. New urban expansion projects, when seeking external conditions for their own survival and development, can maintain convenient collaborative connections with the city center by choosing locations adjacent to it. This allows them to fully utilize the city's infrastructure and social service facilities, thereby reducing investment costs. However, as the city size further expands, the disadvantages of this development model become increasingly apparent. For example, successive ring roads become artificial thresholds for urban expansion, various functional areas become severely mixed, and the city center becomes overburdened, among other issues.

2.3.5 Natural Conditions Factors

This refers to the natural geographical features surrounding a city, such as terrain, geology, hydrology, climate, etc. As fundamental conditions for urban development, they directly affect and even determine, to a certain extent and period, the spatial structure and morphology of urban evolution through various aspects such as city site selection, urban pattern, land potential, functional area organization, engineering networks, green space layout, and the organization of urban landscapes. For instance, the layout of industrial and residential areas should consider wind direction constraints, construction lands should be chosen in areas with high land bearing capacity and good flood prevention conditions, and areas with steep slopes and low bearing capacity should be designated for green spaces. Of course, natural conditions usually do not independently affect urban spatial structure but do so in conjunction with other factors (Liu and Tang 2001). They are a necessary condition but not a sufficient one. As technology advances, the impact of nature on humans may diminish, but it remains one of the important factors influencing urban spatial layout.

2.3.6 Technological Factors

From the perspective of urban development history, technological innovation is an important factor in the evolution and development of urban spatial structure. By transforming production methods, it changes existing production relationships, influencing and promoting changes in people's production and living styles. This is reflected in the evolution and renewal of urban space. Among the most important technological innovations are advancements in transportation and communication technologies.

Transport accessibility plays a decisive role in the evolution of urban space. Different transportation eras have led to different urban spatial structures (Qiu 2002). In the era of walking and horse-drawn carriages, under the influence of "conventional road networks," urban spatial structures often exhibited a single-center concentric circle layout. Later, entering the era of trams and trains, the reach of cities rapidly expanded, prompting urban expansion along transportation routes. With the rise and popularization of private cars, the internal "rapid road system" of cities reduced the dependence of urban residents' living and production activities on city centers, leading to sprawl towards suburban edges. The construction of modern highways and other infrastructure between cities has further led to a decentralization of urban space. Different levels of road systems each carry distinct traffic functions, both connecting and isolating within the city, forming a complex urban road network system. The road network of a city often has long-term stability, and thus forms the "basic framework" of the modern urban layout. On the other hand, advancements in transportation modes also affect the internal regional differentiation of cities. Urban regional differentiation is the result of various economic activities continuously selecting favorable locations (Qiu 2002). The convenience of transportation environments in different urban areas varies significantly, guiding the location choices of businesses and consumers, thereby affecting urban land use. Moreover, a well-developed transportation system is a prerequisite for large-scale urban development (Wu and Ma 1990). Continuous improvements in transportation infrastructure have broken the spatial and temporal constraints on urban residents' travel, leading to large-scale population movements for employment and residence to the urban periphery and outer suburbs. This triggers large-scale development along roads and at the edges, with urban space expanding directionally based on axial characteristics. This expansion increases the spatial distance between different urban areas, leading to regional differentiation on a larger spatial scale (Zhou and Li 1999).

Advancements in communication technology have an increasingly evident impact on the evolution of urban spatial structure. The digitization and networking of urban production and life reduce the limitations of spatial distance on people's activities, enhancing the efficiency and capacity of urban material exchanges. This flexibility in location choice for activities allows businesses to choose lower-cost suburban areas for factory locations, and residents to prefer beautiful suburbs for living, leading to a more dispersed urban spatial pattern and even potentially causing a restructuring of the entire urban spatial structure.

2.3.7 Policy and Institutional Factors

Every stage of a city's genesis and development is closely related to policies and institutions, with urban spatial evolution also bearing a distinct political aspect. In urban management, planning controls and policies for urban construction and development variously promote or restrain the evolution of urban spatial structure. Urban planning, as the primary means of intervening in urban construction, plays an increasingly important role in adjusting and curbing the development of unreasonable urban forms. Its core objective is to address the scale and spatial layout of urban land use. The functional use, development intensity, and environmental requirements of urban parcels all depend on the control requirements of urban planning.

On one hand, central cities, through land expansion, gain broader development space. New urban planning must consider industrial layouts and urban spatial structure layouts on a larger spatial scale. On the other hand, the spatial development of absorbed suburban areas faces both opportunities and challenges. They must first serve as important reception areas for the relocation of populations and industries from central cities. Moreover, as part of a multi-nucleated metropolitan area, these suburban areas often face challenges like diminished subjective initiative in spatial development and inefficiencies due to increased bureaucratic procedures.

The different processes of planning implementation also affect the modes of urban spatial expansion. Although the intervention of city governments in urban development has gradually shifted to indirect regulation, their control remains strong during the process of systemic transition, exerting a significant influence on urban development and land use. Thus, government policies and planning control remain important factors in the development and evolution of urban spatial structure.

2.4 Stages and Forms of Urban Spatial Evolution

Urban development is always in a state of constant change, with its spatial form also undergoing a series of stage-wise evolutionary processes, primarily reflected in the expansion of urban space. Despite the complex and diverse ways of evolution due to differences in the urban development environment, this stage-wise evolutionary process, as a common phenomenon of urban development, has some shared characteristics. In terms of stages of urban development, it generally goes through five phases: point formation, axial expansion, stabilization of extended axes, inward infill, and re-axial extension. Regarding the basic modes of urban spatial form evolution, there are four main types: sprawl growth, contiguous growth, extended axial growth, and leapfrog expansion. Therefore, the evolution of urban space exhibits periodic patterns in the extent of expansion, axial regularity in the direction of expansion, and functional structural patterns in the driving force of expansion (Hu 2002).

The evolution of urban spatial structure is always influenced and guided by two types of forces: unintentional natural growth and development, and intentional human control. The interplay of these two forces constitutes the primary motive power for urban formation, development, and demise. Human intervention in urban evolution has been almost as old as the cities themselves. This intervention is proactive and purposeful, and it can impact the evolutionary process of urban spatial structure in three ways: first, when human organizational power synchronizes with the self-organizing power of urban space, it accelerates urban development; second, it can modify the self-organizing process and direction of space, guiding the evolution of urban spatial structure towards a specific direction; third, it can obstruct or delay the self-organizing process and direction. This entirely depends on the purpose, method, and capability of the active human intervention, and it is directly related to people's values and subjective orientations.

2.5 Significance of Urban Spatial Evolution Simulation Research

China is currently focusing on advancing a new type of urbanization, with a clear mandate to prioritize population urbanization and comprehensively enhance the quality and level of urbanization. As of 2019, China has achieved an urbanization rate of 60% of its permanent population. According to the national goal of essentially achieving modernization by 2035, there is still a need to progress towards a 70% urbanization rate. Unlike the extensive land urbanization of the past, people-centered urbanization will inevitably trigger a socio-economic spatial restructuring of cities. Therefore, simulating the urban spatial evolution process, scientifically and proactively predicting urban spatial development trends, and formulating effective response measures are unavoidable tasks.

Current domestic research on urban space focuses mainly on empirical analysis, with integrated simulation studies being relatively weak. Urban spatial decision-making still awaits strong scientific support. The Land Use-Transport Interaction (LUTI) model provides an important avenue for simulating urban spatial development. However, its application research is primarily conducted in developed countries, often for specific cities, and is seldom commercialized. Moreover, traditional LUTI models face a bottleneck in achieving more refined socio-economic spatial simulations and urgently need new implementation approaches and methods.

This book aims to elucidate the methods of simulating urban spatial evolution, with a focus on explaining the concept, principles, and implementation methods of the LUTI model. Using Beijing as an example, it establishes a dynamic, "activity-based" LUTI model to simulate the evolution process of urban socio-economic activity spaces and conduct scenario experiments. The model represents the development and application of the LUTI model in Chinese cities. This enriches domestic research on urban spatial evolution simulation and supports urban spatial decision-making.

2.6 Chapter Summary

The continuous evolution of urban space is a classic subject of interest in urban geography, with a long history of academic accumulation. This chapter analyzes the concept of urban spatial evolution and briefly reviews classic theories related to urban space. These include Howard's Garden City Theory, Le Corbusier's Modern Urban Theory, Burgess's Concentric Zone Theory, Hoyt's Sector Model, Harris and Ullman's Multiple Nuclei Theory, Christaller's Central Place Theory, the Machu Picchu Charter Thought Theory, Satellite and New Town Development Theories, and the Theory of Organic Dispersion. These classical theories have played a significant role in our understanding of urban space and its evolutionary processes. The chapter also outlines the factors influencing urban spatial evolution, including economic, demographic, cultural factors, urban functions and size, natural conditions, technology, and policy and institutional factors. Urban spatial evolution is influenced by various factors, and these are just some of the more common major influences. Research on the simulation of urban spatial evolution should focus on these main influencing factors.

Building on the review of classic urban space theories and analysis of factors influencing urban spatial evolution, this chapter further analyzes the stages and characteristics of urban spatial evolution, clarifying the significance of simulating the urban spatial evolution process. Urban spatial evolution is a complex process, and sound urban spatial policies are essential to ensure the healthy development of urban spaces. As urban spatial development and evolution do not allow for rehearsals, it is urgent to simulate the urban spatial evolution process to test different policy effects and support decision-making. The Land Use-Transport Interaction (LUTI) model provides an important avenue for simulating urban spatial development. However, its application research is mainly conducted in developed countries. Therefore, conducting application research on the LUTI model in Chinese cities, assisting in Chinese urban spatial decision-making, enriching domestic urban research methods, and promoting the development of the LUTI model is of significant importance.

References

Butlin FM, Howard E (1899) Tomorrow: a peaceful path to real reform. Econ J 9:71–72

Christaller W (1966) Central place in Southern Germany. Baskin C (trans) Prentice Hall, London

Feng J, Zhou Y (2002) The spatial pedistribution on and suburbanization of urban population Hangzhou city. Urban Plan 01:58–65

Fu L (2012) Spatial structure and its evolution in metropolises during transition. China Architecture & Building Press, Beijing

Guo H, Jiang M (2002) Urban spatial economics. Economics Science Press, Beijing

Hou R, Xi B (2004) A study on the evolutionary characteristics of urban space in the condition of information technology. Acad Exchange 2:121–125

Hou R, Zhang B (2006) Logistics facility layout and planning based on urban spatial evolution. Urban Problems 4:32–35

Hu J (1995) Chinese cities: patterns and evolution. China Architecture & Building Press, Beijing

Hu H (2002) On the evolution and tendency of urban spatial development with Changshu city in Jiangsu province as the case. Urban Planning 4:64–68

Huang Y (2002) Urban space theory and spatial analysis. Southeast University Press, Nanjing

Liu H, Tang S (2001) Analysis of the urban-rural dual complexity of small towns. Small Town Constr 03:18–21

Qiu J (2002) The impact of transportation progress on urban spatial structure and urban planning. Planner 07:67–69

Sun S (1997) Philosophy of urban planning. China Architecture & Building Press, Beijing

Tang Z (1997) Theories and methods in western urban spatial structure research. Urban Planning Forum 6:1–11

Tang J (2016) Evolution process and dynamic simulation of urban spatial evolution in Lanzhou based on multi-agent model. Northwest University, Xi'an

Tao S, Chen W (2001) Exploring the evolution of urban form and cultural charm of Shanghai. Urban Plan 01:74–76

Wu, Ma Q (1990) Mechanism analysis of spatial structure evolution in urban edge areas. Urban Plann (2):38–42, 64
Yang R, Zhang X (1997) Research on the dynamics and patterns of urban spatial expansion. Regional Res Dev (02):2–5+22
Zhang T (2001) The urban restructuring of Chinese cities in 1990s and its dynamic mechanism. Urban Planning Forum 7:7–14
Zheng X, L Lin (2002) A review of studies on the urban morphology since 1990s (07):59–64+92
Zhou X, Li C (1999) RA study of the traffic development in the metropoils. Planner 01:96–99
Zhu Y, Yao S (1999) Study on urban spatial evolution in the Yangtze River Region of Jiangsu and Anhui provinces. Econ Geogr 3:48–53
Zhu Y, Yao s, Li Y (2000) On the urban spatial evolution in the process of urbanization in China. Geogr Territorial Res 2:12–16

Chapter 3
Urban Spatial Analysis and Simulation Methods

Urban space, as a classic subject of study in human geography, has long been a focus of academic attention and has accumulated a vast array of research methods, commonly referred to as urban models or even simulation methods. There are significant differences between these models. Based on their functions, the author categorizes them into two main types: analysis models and simulation models.

Urban analysis models are typically used to assist researchers in understanding urban phenomena, describing the interactions between various elements of a city, and are often specific mathematical models. Urban analysis models help us recognize and describe the laws of urban spatial development, which is a prerequisite for simulation. Simulation models, on the other hand, fit the process of urban spatial development. They are abstract representations of the development process of phenomena, with their primary function being to predict urban spatial development trends. The implementation of urban spatial simulation models usually involves multiple mathematical models and can be achieved using different mathematical models, representing a further deduction of the laws of urban spatial development.

3.1 Urban Spatial Analysis Model

3.1.1 Overview of Urban Spatial Analysis Models

Urban analysis models are those that help us understand and describe the laws of urban spatial development. They can be broadly divided into two categories: statistical analysis models and spatial analysis models.

Statistical analysis models, such as variance analysis, correlation analysis, regression analysis, cluster analysis, principal component analysis, and input–output analysis, help in deciphering the quantitative relationships among various urban elements.

F. Niu, *Urban Spatial Evolution Simulation*, Urban Sustainability,
https://doi.org/10.1007/978-981-97-3481-8_3

Spatial analysis models focus on the interrelationships within urban spaces, including models like spatial diffusion, distance decay, gravity models, and potential models.

Cities are spatial entities, and the development of urban spatial analysis models is a part of the evolution of urban space research. In 1933, Professor Colby, an urban geographer from the University of Chicago, proposed the concept of centripetal and centrifugal forces to describe changes in urban land use and regional differentiation (Zhou and Chen 2003; Zhou et al. 1999). American scholar William J. Reilly, after surveying 150 cities in the United States and combining the law of universal gravitation, summarized the relationship between urban population and retail gravity in 1931, proposing the "Law of Retail Gravitation" (Reilly 1959). This law suggests that a city's attractiveness to surrounding areas is directly proportional to its size and inversely proportional to the distance between them, laying the foundation for quantitative analysis of regional spatial connections. Subsequently, P. D. Converse, building on Reilly's theory, introduced the concept of the breaking point in 1949 (Converse 1949), defining the boundary of urban connections and using survey data on residents' shopping travel to validate and calibrate the model. Zipf further developed Reilly's law and established the theoretical basis for spatial interactions within urban systems (Zipf 1946). In 1999, Chen Shunqing proposed the concept model of urban development centripetal force, centrifugal force, and friction based on research viewpoints on urban growth and land appreciation, which could initially reflect the formation process and development changes of urban organizational structure (Chen 1999a). In 1953, Swedish scholar T. Hagerstrand first addressed the issue of urban spatial diffusion in his article "Innovation Diffusion as a Spatial Process," identifying forms of diffusion such as wave, radiation, hierarchical, and leapfrogging, and initiated research into spatial diffusion (Hagerstrand 1967).

There are numerous models for urban spatial analysis, too many to exhaustively list. In summary, these models focus on summarizing the characteristics and laws of urban spatial development. Below, we provide examples of two relatively complex analysis model applications: the difference-in-differences model and the complex network analysis model.

3.1.2 Application of the Difference-in-Differences Model—Spillover Effects of High-Speed Rail Stations in China and Their Spatial Variations

1. **Research Question**

China is actively constructing a high-speed rail (HSR) network. In August 2008, the Beijing-Tianjin Intercity Railway, the first high-speed rail in China with a design speed of 350 km per hour and independent intellectual property rights, began operations, marking China's official entry into the high-speed rail era. By 2018, China's high-speed rail mileage reached 29,000 km, accounting for two-thirds of the world's total high-speed rail mileage, and the annual passenger volume increased from 7.34

million in 2008 to 3.37 billion. China has become the country with the longest high-speed rail mileage and the most complex operation network in the world (Hu et al. 2019). Focusing on the impact of high-speed rail on urban economic development, scholars have conducted numerous studies. However, due to different case areas, research methods, and periods studied, various conclusions have been reached. One view is that high-speed rail can enhance intercity connections and further strengthen the trickle-down effect of knowledge and technology from central cities (Albalate and Bel 2012), spreading quality resources and elements to surrounding areas, thereby driving economic growth, narrowing regional disparities, and achieving coordinated regional economic development (Ahlfeldt and Feddersen 2017; Campos and Rus 2009; Wang and Ni 2016; Zheng and Kahn 2013). Another view suggests that the operation of high-speed rail is more advantageous for cities with strong competitiveness. For cities with relatively weaker competitiveness, it merely creates opportunities to improve their status, with potential negative effects leading to economic activity outflows. The final outcome depends on the combined effect of spillovers and echo effects (Ahlfeldt and Feddersen 2017; Cui et al. 2018; He and Huang 2020; Pol 2003). For example, Kim found that after the expansion of high-speed rail in 2010–2011, spatial equity in South Korea was deteriorating, with the benefits of improved transportation accessibility concentrated in the already developed Seoul region (Huang and Han 2019); Givoni's study concluded that high-speed rail might promote regional economic growth in the short term, but over the long term, the economic growth rate of areas on the periphery of the high-speed rail network would decline (Givoni 2006).

Despite varying conclusions, the operation of high-speed rail stations in China will undoubtedly enhance the accessibility of surrounding areas, inducing economic activity aggregation, reflected in increased land use development intensity and rising real estate prices (Diao 2018)—the high-speed rail station spillover effects that this article focuses on. Many local governments view high-speed rail stations as engines of economic development, investing heavily in areas around high-speed rail stations, hoping to leverage the high-speed rail dividend to revitalize the local economy and aid the urbanization process. The impact of high-speed rail on regional economic development has become a hot topic in academia. Current studies mostly use administrative divisions as observation units (Yu and Zhao 2019; Zheng et al. 2019;) and treat the spatial spillover effects of high-speed rail stations on surrounding cities as homogeneous and undifferentiated (Banister and Berechman 2001). However, cities are geographical spatial entities with complex network structures and spatial hierarchies. The impact of high-speed rail on urban development is usually concentrated around high-speed rail station areas (Ureña et al. 2009; Zheng et al. 2019). Some high-speed rail stations far from city centers tend to have an impact radius that crosses administrative boundaries, making traditional statistical indicators based on administrative divisions (such as GDP) inadequate to reflect the true spillover effects of high-speed rail stations (Liu and Yang 2009; Xu et al. 2015a).

When statistics based on administrative divisions fail to describe the microstructure of cities (Liu and Yang 2009; Xu et al. 2015a), nighttime light indices offer a new approach. Previous studies have shown that urban nighttime light is highly consistent with the distribution of economic activities (Elvidge et al. 2009). Nighttime light data can be gridded at different scales and has been widely used in transportation and urban spatial structure research (Amaral et al. 2006; Baum-Snow et al. 2017; Elvidge et al. 1997a, b; Ghosh et al. 2010; Gonzalez-Navarro and Turner 2018; Wang et al. 2020). Based on nighttime light data and using the difference-in-differences model, we measure the spatial spillover effects of high-speed rail stations, exploring their spatial differentiation patterns and influencing factors, hoping to provide references for the placement of high-speed rail stations in China and planning of new new high-speed rail cities.

2. **Research Method—Difference-in-Differences Model**

The changes in economic activities around high-speed rail (HSR) stations stem from two sources: one is the effect generated by the construction and operation of the HSR station, known as the spillover effect, or "policy effect," and the other is the natural growth brought about by the city's own expansion (Givoni 2006), referred to as the time effect. Therefore, to analyze the spillover effect of HSR stations, it is necessary to eliminate the time effect and isolate the spillover effect from changes in economic activities around the station. For this purpose, the opening of high-speed rail is regarded as a quasi-natural experiment, using the differences in the cities where HSR stations are located and the timing of their opening. We employ a progressive difference-in-differences method (multiple time point DID) to calculate the spillover effect of HSR stations. DID can control general factors affecting economic development and unobservable factors that do not change over time. It is effective in identifying policy effects, reducing errors, and enhancing the explanatory power of the model (Beck et al. 2010), and is widely used in research on high-speed rail and urban expansion (Ahlfeldt and Feddersen 2017; Urena et al. 2009).

The difference-in-differences method requires a control area (an area not affected by the HSR station) to be set for the study area (the area around the HSR station, i.e., the area influenced by the HSR station's spillover effect), ensuring that both groups have the same development trend before the policy impact (parallel trend). For this purpose, we selected a 4-km radius around the station as the study area to calculate the spillover effect of HSR stations. The area corresponding to this radius is approximately 50 km^2, roughly the average size of urban areas in China (Diao 2018). At the same time, an annular area of 4–8 km around the HSR station is selected as the control area. The selection of the experimental and control areas must meet parallelism and robustness tests (see below). We extracted the average light intensity (Average Digital Number, ADN) for both the experimental and control areas around the HSR stations. The difference-in-differences model used is as shown in Eq. (3.1) (Diao 2018; Lin 2017; Liu and Li 2017; Shao et al. 2017).

$$Y_{\mathrm{it}} = \alpha + \beta HSR_{it} + X_{it}\gamma + \mu_i + \lambda_i + \varepsilon_{it} \tag{3.1}$$

where α. is a constant and Y_{it}. is the intensity of economic activity around site i (study or control area) in year t, characterized by ADN, which has the advantage of overcoming the "endogeneity problem" due to bi-directional causality in the production function; The core explanatory variable HSRit is a dummy variable characterizing whether or not the study area is open to high-speed railway, and the HSR of the study area in the year when the high-speed railway opens and in each year after that takes the value of 1, otherwise is 0; Xit is a vector including site location variables, socio-economic variables, which is used to control the influence of other factors on the economic development of the city; μ_i is an individual fixed effect, which characterizes that the sample does not change over time; λ_i. is a time fixed effect; ε_{it}. is a random error term; the statistic β is used to measure the net spillovers generated after the commissioning of the high-speed railway station, and if the net spillover effect is significant, it indicates that the opening of the high-speed railway has significantly stimulated economic activities in the neighboring regions.

3. **Research Data**

(1) **Nighttime Light Data**

The most widely used nighttime light data currently are the Defense Meteorological Satellite Program's Operational Linescan System (DMSP/OLS) data (1992–2013) and the Visible Infrared Imaging Radiometer Suite (VIIRS) data onboard the next-generation polar-orbiting satellites (NPP/VIIRS) (starting in 2012) (Li et al. 2018). DMSP/OLS nighttime light data include 34 annual composite images from 22 years between 1992 and 2013, captured by six different sensors. The pixel values, DN (Digital Number), represent the light intensity and are relative brightness values that have not been calibrated on-board the satellite. Due to differences in sensors, satellite data from different periods are not comparable (Cao et al. 2015; Elvidge et al. 2009). NPP/VIIRS data consist of 69 monthly composite images from 2012 to 2017. The NPP/VIIRS sensor added an in-flight correction feature, and the pixel DN values are absolute brightness values after radiometric calibration, making the series of image data comparable. Table 3.1 shows the different satellite data numbers and their operational periods.

Combined with the history of China's high-speed railway development, this section adopts the nighttime lighting data from 2004 to 2017, which spans two generations of satellites, and thus requires two aspects of calibration: first, DMSP/OLS time-series data calibration to realize the comparability of the nighttime lighting data in each year; second, the connection of data from two generations of satellites to realize the comparability of the two. The pixel-by-pixel calibration method can be

Table 3.1 Different satellite data numbers and operation periods

Satellite	Operation period
F10	1992–1994
F12	1994–1999
F14	1997–2002
F15	2000–2007
F16	2004–2009
F18	2010–2013
VIIRS	2012–2017

used to calibrate the DMSP/OLS sequence image data: first, select the image of the unchanged area in a specific year as a reference; then, calibrate the nighttime light data by regression to establish a functional relationship between the pixel values of triginal image and the calibrated pixel values (Zhao et al. 2019), as shown in Eq. (3.2).

$$DNc = a * DN^2 + b * DN + c \tag{3.2}$$

where DN and DN_c are the pixel values of the pixels in the reference area before and after correction, respectively; a, b, c are parameters. The regression results of the parameters are shown in Table 3.2.

Figure 3.1 shows the statistical results of the Average Digital Number (ADN) per unit area within the reference area before and after continuity correction of the DMSP/OLS sequence image data, which shows that the calibrated sequence of ADN per unit area is more continuous.

The DMSP/OLS and NPP/VIIRS sequence image data overlap in 2012 and 2013. Focusing on calibrating the temporal and spatial overlap of the two sets of sequence image data, there are two commonly used calibration methods: the first one is the image element calibration method (Lv et al. 2020; Zhao et al. 2019), which selects a specific region in the overlapping years of the two sets of sequence image data, extracts DN, and establishes a functional relationship between the DN of the two sets of sequence image data; the second is the pseudon invariant features method (PIFs; Jeswani et al. 2017), which extracts pseudo invariant feature areas using a combination of Getis-ord Gi* statistics and coefficients of variation (CV) over 3 × 3 and 5 × 5 pixel windows, and constructs the fitted model using a sufficient number of pseudo invariant feature areas. The optimal calibration function using the pixel calibration method is $Y = 152.72x^{0.7204}(R^2 = 0.8)$, where X and Y are the DN of the DMSP/OLS and NPP/VIIRS sequential image data, respectively. The pseudo invariant feature method yields the optimal calibration function as $Y = 38.67 \ln X + 12.07(R^2 = 0.8)$, where X and Y are the comparable DN, which is calibrated DN

Table 3.2 Regression results of model parameters for multi-year image DN correction

Satellite	Year	a	b	c	R^2
F16	2004	−0.0016	1.1666	1.0701	0.8775
	2005	−0.0026	1.442	0.1668	0.9016
	2006	−0.0055	1.4985	0.0195	0.9083
	2007	0.0000	1.0000	0.0000	1.0000
	2008	0.0071	0.6311	1.3309	0.9000
	2009	0.0079	0.4576	1.9299	0.8909
F18	2010	0.0085	0.4456	2.0004	0.8229
	2011	0.0084	0.481	2.4218	0.8412
	2012	0.0086	0.5017	2.7798	0.8816
	2013	0.0084	0.4833	2.5755	0.8266

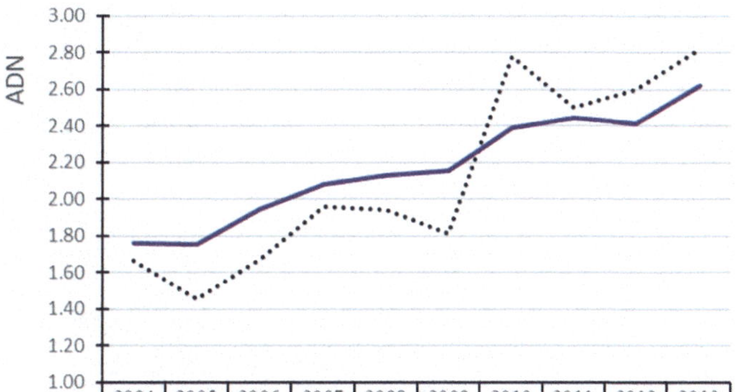

	2004	2005	2006	2007	2008	2009	2010	2011	2012	2013
••• Before calibration	1.66	1.45	1.67	1.96	1.94	1.81	2.78	2.50	2.60	2.82
— After calibration	1.76	1.75	1.95	2.08	2.13	2.15	2.39	2.44	2.41	2.62

Fig. 3.1 Comparison of ADN per unit area in the reference area before and after continuity correction of DMSP/OLS sequence imagery data (2004–2013)

of VIRS. Both methods can achieve good results. Comparing the fitting coefficients and imaging effects of the two methods, the pseudo invariant feature method, which has more continuous calibration results, is used in this section to calibrate the data and realize the comparability of the two sets of serial image data.

(2) **Railway Transportation and Socioeconomic Data**

This study involves high-speed trains that start with the letters "C," "D," and "G" (Jiao et al. 2016). According to the information published by the China State Railway Group Co., Ltd., as of 2017, there were 527 high-speed railway stations in 180 cities across the country, as shown in Fig. 3.2. It is evident that the distribution of stations is denser along the coast and sparser in inland areas, especially in remote western regions. The number of cities that have opened high-speed railways in the eastern, central, western, and northeastern regions of China are 67, 60, 39, and 14, respectively, accounting for 37.2, 33.3, 21.7, and 7.8% of the total. High-speed railway stations are categorized according to their construction model into newly built stations and upgraded stations; details are provided in the results analysis section.

To analyze the impact of urban socioeconomic factors on the spillover effect of high-speed railway (HSR) stations, this study selects indicators at the city level including GDP, total population, fixed asset investment, industrial structure, and the number of employees in secondary and tertiary industries. These indicators are all important in representing the socioeconomic status of a city (Zhang 2017). Station-level indicators include the number of high-speed rail stations in the same city, the presence or absence of an airport in the city, the distance of the high-speed rail station from a major city, and the number of trains operating at the station. The number of high-speed rail stations in the same city and the presence or absence of an airport are used to consider the competitive and cooperative relationship between high-speed rail stations and between stations and airports, a current research hotspot.

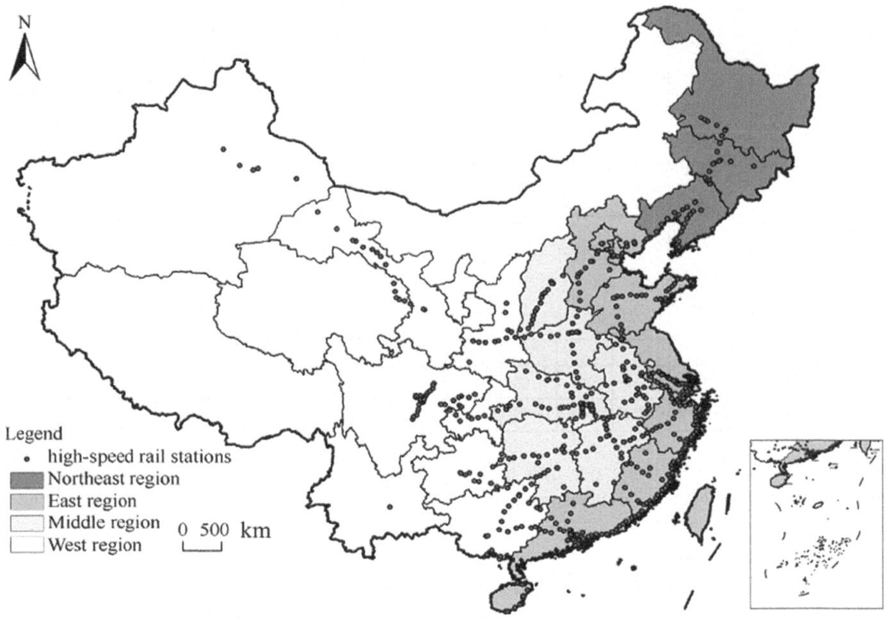

Fig. 3.2 Distribution of high-speed railway stations in China. *Note* Data for Taiwan unavailable

In addition, existing studies show that the distance between a city and a major city is an important factor affecting high-speed rail benefits (Zheng and Kahn 2013). Therefore, this study includes the distance variable between high-speed rail stations and major cities, using the shortest high-speed rail distance to Beijing, Shanghai, and Guangzhou. The number of operating trains is used to represent the level of the station, serving as a proxy for station passenger throughput (there are no publicly available statistics on station passenger throughput). After excluding stations with severely missing data, 520 stations remain, with corresponding indicator values as shown in Table 3.3. ADN represents the intensity economic activity in the area around the station, characterized by the nighttime light intensity per unit area. This study interprets the light intensity values around all the aforementioned stations, corresponding to 180 cities (see below for the research method).

4. **Results Analysis**

(1) **Overall Spillover Effect of High-Speed Railway Stations Nationwide**

In this study, the intensity of economic activity (ADN) around high-speed rail (HSR) stations is used as the dependent variable. Initially, a mixed regression of the full sample for the experimental group (sample within a 4-km radius around the station) is conducted using the least squares method, and the results are shown in column (1) of Table 3.4. The HSR coefficient is 0.4011 and is significant at the 1% level, indicating that the opening of high-speed rail promotes the enhancement of surrounding economic activities. Column (2) shows the results of the difference-in-differences

Table 3.3 Statistical description of key indicators for high speed railway stations

Indicator	Type	Description	(1)	(2)	(3)	(4)	(5)
			Sample size	Average value	Standard deviation	Minimum	Maximum
ADN	Intensity of economic activity in the area surrounding the site	Average nighttime light brightness values	14,560	55.74	29.13	1.25	264.50
HSR	virtual variable	Whether or not to open high speed railway	14,560	0.224	0.417	0	1
GDP	Urban socio-economic variables	Gross regional product (billion yuan)	7127	3010	3930	39.357	30,000
Cap	Urban socio-economic variables	Investment in fixed assets (billions yuan)	7258	877.14	1352.49	6.66	9920.37
Pop	Urban socio-economic variables	Total population (million people)	7253	555.78	435.59	16.76	3392
Empl	Urban socio-economic variables	Number of employees in secondary and tertiary industries (million people)	7084	455.66	5663.44	4.03	123,009.3
Ind_Stru	Urban socio-economic variables	Industrial structure (ratio of value added of secondary and tertiary industries)	7236	1.29	0.52	0.23	6.73

(continued)

Table 3.3 (continued)

Indicator	Type	Description	(1)	(2)	(3)	(4)	(5)
			Sample size	Average value	Standard deviation	Minimum	Maximum
Sta_ Num	Site location variables	Number of high-speed railway stations in the same city	180	4.623	3.510	1	18
Air	Site location variables	Whether or not to have airport in the same city	180	0.438	0.496	0	1
Dis_ Meg	Site location variables	Distance between high speed railway station and metropolis (kilometers)	177	1154	2121	34.83	10,000
Train_ num	Site level variables	Number of trains operating at high-speed railway stations	520	46.04	63.60	1	503

Note The sample size of the city socio-economic variable is inconsistent because of missing variables; the sample size of the Dis_Meg variable excludes three cities: Beijing, Shanghai, and Guangzhou

model regression controlling for "time fixed effects" and "individual fixed effects," where the high-speed rail effect HSR coefficient is 0.0675, significant at the 5% level. This again indicates that the operation of high-speed rail stations significantly enhances the intensity of economic activities in the surrounding area. Comparing the results in columns (1) and (2), the HSR coefficient in the difference-in-differences regression is smaller, but the fit is better (higher R2), suggesting that ordinary least squares regression significantly overestimates the spillover effect of high-speed rail stations. This aligns with our expectations that the changes in economic activities around high-speed rail stations are not entirely due to the spillover effect, confirming the effectiveness and scientific validity of using the difference-in-differences model to evaluate the high-speed rail effect.

Column (3) shows the two-way fixed-effect regression results including socioeconomic variables like GDP and population. The results show that the estimated HSR effect remains significant at the 5% level, with a slight reduction in absolute value compared to column (2), indicating that regional socioeconomic conditions

Table 3.4 Full sample regression results

Dependent variable: ln(ADN)	(1) Full	(2) Full	(3) Full	(4) Full	(5) New station	(6) Upgraded station
HSR	0.4011*** (0.014)	0.0675** (2.543)	0.0669** (2.490)	0.0474* (1.947)	0.0476* (1.764)	0.0437 (0.781)
				−0.0102*	−0.0136**	0.000580
Sta_Num				(−1.810)	(−2.208)	(0.0404)
				−0.118***	−0.0694	−0.261***
Air				(−2.863)	(−1.589)	(−2.672)
				−0.000608	0.00993	−0.0551
ln(Dis_ Meg)				(−0.0333)	(0.510)	(−1.092)
				0.00430	−0.00131	0.00821
ln(Train_ num)				(0.327)	(−0.0883)	(0.285)
			0.0329	0.00709	0.0120	−0.0539
ln(GDP)			(0.460)	(0.186)	(0.291)	(−0.504)
			0.0107	0.0106	0.00219	0.0443
ln(Cap)			(0.532)	(0.595)	(0.111)	(1.045)
			0.0113	−0.0459	−0.0257	−0.110
ln(Pop)			(0.0889)	(−1.265)	(−0.675)	(−1.122)
			−0.0396	−0.0455*	−0.0422	−0.0355
Ind_Stru			(−1.483)	(−1.919)	(−1.472)	(−0.834)
			0.0135	0.0361**	0.0344**	0.0493
ln(Empl)			(0.357)	(2.146)	(2.022)	(0.697)
	3.591***	3.558***	2.927**	3.708***	3.493***	5.202***
Constant	(0.010)	(0.0157)	(2.353)	(7.300)	(6.440)	(3.464)
Individual fixed effect	–	Yes	Yes	No	No	No
Time fixed effect	No	Yes	Yes	Yes	Yes	Yes
Sample size	7280	14,560	14,084	13,440	10,896	2544
R^2	0.100	0.303	0.305	0.306	0.300	0.329

Note 1. Due to the absence of data for some stations or cities, the number of samples included in the regression analysis varies across groups (the number of samples in column (4) is less than in column (3) because some high-speed rail stations that do not have connections to the three major metropolises of Beijing, Shanghai, and Guangzhou were excluded). 2. Numbers in parentheses are the t-test values corresponding to the results, similarly the same in the later text. 3. *** indicates a 99% significance level, ** indicates a 95% significance level, and * indicates a 90% significance level, similarly the same in the later text somewhat affect the measurement of light intensity, but the main spillover effect of high-speed rail stations comes from their operation. Column (4) presents the full-sample regression results with all control variables included. The results show that the economic activity intensity in the experimental area increased by about 4.7% compared to the control area, providing strong evidence of a significant spillover effect from high-speed rail stations. In the regression results, the coefficient of the number of stations in the same city (Sta_Num) is negative and significant, suggesting that the agglomeration effect of stations on surrounding economic activities weakens as the number of high-speed rail stations in the same city increases. A possible reason is the diversion of people and resources flowing into high-speed rail cities among various stations, reducing the impact of individual stations. The estimation result for airports (Air) is significantly negative and has a large absolute value, indicating that the spillover effect of high-speed rail stations is greatly reduced in cities with airports, suggesting a clear competitive relationship between high-speed rail and civil aviation. Observing other economic statistical variables, the estimated coefficient for employment opportunities (Empl) is consistently positive, indicating that cities with more job opportunities have greater development potential and higher transportation demand, promoting the agglomeration of economic activities around high-speed rail stations. The coefficient for industrial structure (Ind_Stru) is negative, indicating that areas with a smaller proportion of secondary industries or a larger proportion of tertiary industries receive a higher spillover effect from high-speed rail stations. This is because the tertiary industry is more labor-intensive than the secondary industry and requires more labor. Therefore, the further development of the tertiary industry will absorb more employment, generating transportation demand. This result also indicates that the development of the service sector will further enhance the economic activity intensity around high-speed rail stations.

According to the construction model of high-speed rail (HSR) stations, they can be divided into upgraded and newly built stations (accounting for 19% and 81% respectively). The reason for this classification is that upgraded stations are usually located in built-up areas where land acquisition and development costs are substantial. Therefore, upgrading existing stations can improve land use efficiency and achieve natural synergy between the high-speed rail network and the original urban transportation network (Diao 2018). In contrast, newly built stations are often located in suburban areas or urban–rural boundaries, where land development is relatively easier. Hence, the spillover effects of these two types of HSR stations are expected to be significantly different. As shown in columns (5) and (6) of Table 3.4, compared to upgraded stations, newly built high-speed rail stations have a more significant spillover effect. This aligns with our understanding that newly built stations are usually located in suburban areas, surrounded by agricultural land with weaker light intensity, making them easier to develop. Upgraded stations, typically situated in built-up areas or closer to city centers, already have substantial economic activities around them before the opening, making large-scale changes difficult.

The variable number of stations in the same city (Sta_num) is significantly negative in the newly built group, indicating that the increase in stations has a more significant impact on newly built stations, dispersing the spillover effect and not conducive to the agglomeration of economic activities around new stations. The competitive relationship between airports and high-speed rail is more common in the upgraded group (with a significantly negative Air coefficient), as in cities with large population flows and strong economic foundations, there is a more frequent choice of travel modes. Some scholars have pointed out that cities where high-speed rail competes with the civil aviation network were initially distributed in the eastern and central regions, with the competitive network showing a decreasing spatial distribution from east to west (Qin 2016), which is consistent with the findings of this study.

(2) **Regional Differences in the Spillover Effects of High-Speed Railway Stations**

Based on the division of China into four major regions: East, Central, West, and Northeast, a grouped regression analysis was conducted for high-speed railway (HSR) stations nationwide. The results, as shown in Table 3.5, indicate that the coefficient of the dummy variable HSR is significantly positive in the Eastern region, while it is not significant in the Central, Western, and Northeastern regions. This suggests that compared to the Central, Western, and Northeastern regions, cities in the Eastern region have gained more significant benefits from high-speed rail, indicating that high-speed rail construction is more advantageous for economically developed areas.

(3) **Inter-City Differences in the Spillover Effects of High-Speed Railway Stations**

According to the "Notice on Adjusting Urban Size Division Standards" issued by the State Council, cities with high-speed railway (HSR) stations were classified into four categories based on the permanent resident population at the end of 2017: Megacities (population over 10 million), Very Large Cities (population between 5–10 million), Large Cities (population between 1–5 million), and Medium and Small Cities (population under 1 million). A grouped regression analysis was then

Table 3.5 Regional differences in spatial spillover effects of high-speed railway stations

Dependent variable ln(ADN)	Geographical location			
	Eastern	Central	Western	Northeastern
	0.0877**	0.0365	−0.0153	0.0447
HSR	(2.449)	(0.810)	(−0.290)	(0.407)
Site location variable and Urban socio-economic variables	Yes	Yes	Yes	Yes
Time fixed effect	Yes	Yes	Yes	Yes
Individual fixed effect	No	No	No	No
Sample size	5788	4142	2662	848
R^2	0.286	0.329	0.325	0.348

Note ** indicates a 95% significance level

conducted for these categories, with results shown in Table 3.6. It is observed that only the Very Large Cities category shows a significant positive spillover effect of HSR stations. This indicates that although large cities (usually more developed) can gain more benefits from high-speed rail, the spillover effect of HSR stations decreases when the population exceeds a certain size.

One possible reason is that HSR stations in megacities are typically upgraded stations, where land use is already established and difficult to change significantly. Another potential reason is the diverse industries in megacities, where the introduction of high-speed rail does not impact their existing urban development trajectory.

(4) **The Influence of Metropolises on the Spillover Effect of High-Speed Railway Stations**

The opening of high-speed rail (HSR) enhances inter-city connections, allowing residents to enjoy the diverse products and services and higher wages offered by metropolises, while also living in nearby second and third-tier cities to avoid high living costs and adverse environmental conditions (Shi et al. 2018). Consequently, accessibility to metropolises is an important factor affecting a city's high-speed rail benefits. Based on the shortest distance between high-speed rail stations and the three major cities of Beijing, Shanghai, and Guangzhou, stations are divided into three groups: a short-distance group (less than 500 km), a medium-distance group (500–1000 km), and a long-distance group (more than 1000 km or cities that cannot reach Beijing, Shanghai, and Guangzhou). The analysis results are shown in Table 3.6. The results indicate that only the short-distance group exhibits a significant spillover effect

Table 3.6 Intercity differences in spatial spillover effects of high-speed rail stations

Dependent variable ln (ADN)	City scale				Distance from the metropoils		
	Megacity	Mega-city	Large city	Small and medium-sized city	Near distance	Medium distance	Far distance
	−0.0383	0.0888**	0.0537	−0.00321	0.102***	0.0489	−0.0540
HSR	(−0.568)	(2.211)	(1.575)	(−0.0134)	(2.690)	(1.288)	(−0.961)
Site location variable and Urban socio-economic variables	Yes	Yes	Yes	Yes	Yes	Yes	Yes
Time fixed effect	Yes	Yes	Yes	Yes	Yes	Yes	Yes
Individual fixed effect	No	No	No	No	No	No	No
Sample size	1902	4762	6540	236	5000	5954	2486
R^2	0.389	0.284	0.300	0.428	0.276	0.329	0.342

Note *** indicates a 99% significance level, and ** indicates a 95% significance level of high-speed rail stations, and as the distance increases, the HSR coefficient becomes insignificant or even negative. This suggests that cities closer to metropolises receive greater high-speed rail benefits, while those farther away experience a decline in high-speed rail benefits, and may even suffer from a siphoning effect.

5. **Model Verification**
(1) **Robustness Test**

To ensure the robustness of the model evaluation results, which might be influenced by the threshold setting, further robustness tests were conducted. Since the economic development level of different cities and the location of stations can affect the spillover effect of high-speed rail (HSR) stations, we varied the distance threshold between the experimental and control groups to test the model's robustness. The results are shown in Table 3.7. In columns 2 and 3, the experimental groups are defined as areas within 0–3 and 0–5 km around the HSR stations, respectively, while the control groups are defined as areas within 3–8 and 5–8 km around the HSR stations. The results are generally consistent with those of the main regression (column 1), indicating that the model evaluation results are not sensitive to the choice of threshold and confirming the robustness of the model.

(2) **Falsification Test**

The premise of using the DID model is that there is no systematic difference between the test group and the control group before the opening of the high-speed railway, i.e., the average nighttime light brightness of the two groups has the same changing trend (Niu and Xin 2021). To prove that the parallel trend hypothesis is valid, this section carries out a falsification check using the event analysis method, which tests the policy effect by setting multiple high-speed railway opening years (Lin 2017; Qin 2016). Assuming that the actual opening year of the high-speed railway is T, this study sets T − 4 ~ T + 4 as the opening years for the model test, and the corresponding model is shown in Eq. (3.3). In the formula, L denotes the assumed opening year of high-speed railway, and other variables are the same as Eq. (3.1).

$$Y_{it}=\alpha + \sum_{L-T\neq -2} \beta_L * \mathrm{HSR}_{it} + Xit * \gamma + \mu_i + \lambda_t + \varepsilon_{it} \tag{3.3}$$

Table 3.7 Robustness tests for changing distance thresholds

Dependent variable: ln(ADN)	(1)	(2)	(3)
Distance thresholds (km)	4	3	5
	0.0474*	0.0557**	0.0392*
DID	(1.947)	(0.758)	(0.561)
Site location variable and Urban socio-economic variables	Yes	Yes	Yes
Time fixed effect	Yes	Yes	Yes
Individual fixed effect	No	No	No
Sample size	13,440	13,440	13,440

Note ** indicates a 95% significance level, and * indicates a 90% significance level

The regression analysis using Eq. (3.3) yields the value of parameter β as shown in Fig. 3.3. In Fig. 3.3, the horizontal axis indicates the relative time of the opening of the high-speed railway, for example, −3 means 3 years before the opening of the high-speed railway (L = −3), 0 means the year of the opening of the high-speed railway (L = 0), and 3 means 3 years after the opening of the high-speed railway (L = 3). From Fig. 3.3, it can be seen that the estimated values in the first year, the first 3 years and the first 4 years before the opening of high-speed railway are around 0 [the first two years are used as the base period of the model (Beck et al. 2010)], and the 95% confidence intervals also contain 0, which indicates that the differences in spillover effects of the high-speed railway stations can not be rejected as the original hypothesis of 0, and proves that the assumption of the parallel trend is valid. In addition, Fig. 3.3 shows that HSR is significantly positive in the current period and lagged periods 1, 2, 3 and 4, indicating that the spillover effect of HSR stations has a time lag, which is consistent with the findings of previous scholars (Chang and Murakami 2019).

6. **Conclusion**

China is gradually entering the era of high-speed rail (HSR), and many cities hope to stimulate economic growth through high-speed rail construction, launching various high-speed rail new town projects. However, the effect of high-speed rail varies between cities. Fully understanding the impact of high-speed rail construction on urban development is crucial for the placement of high-speed rail stations, planning of high-speed rail new towns, and coordination of regional economic development. This paper uses nighttime light data and a difference-in-differences approach to

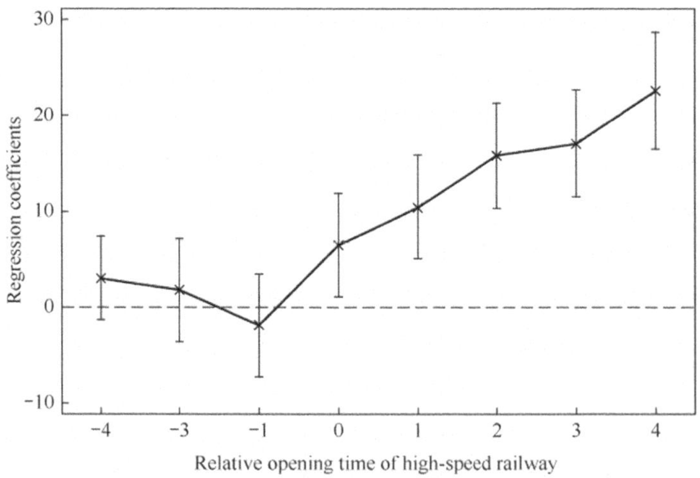

Fig. 3.3 Parallel trends test. *Note* The cross symbols in the graph represent the estimated coefficients, and the short vertical lines represent the 95% confidence interval of the estimated coefficients

analyze the spatial spillover effects of high-speed rail stations in China. The study finds:

(1) Nationwide, the operation of high-speed rail stations has increased the intensity of economic activities in the surrounding areas by about 4.7%. Compared to the Central, Western, and Northeastern regions, cities in the Eastern region exhibit more significant spillover effects from high-speed rail stations.
(2) In terms of population and industry, there is a nonlinear relationship between urban population size and the spillover effect of high-speed rail stations. Once the population exceeds a certain size, the spillover effect is no longer significant, with very large cities exhibiting the most significant spillover effects. The development of the tertiary industry in cities can enhance the spillover effect of high-speed rail stations.
(3) An increase in the number of high-speed rail stations in the same city weakens the spillover effect, especially for newly built stations. The presence of airports also reduces the spillover effect of high-speed rail stations, a competitive relationship more evident in economically developed cities. Cities closer to metropolises have more significant spillover effects from high-speed rail stations, which become less significant or even negative as the distance increases.

Although these insights are preliminary, they are valuable for the placement of high-speed rail stations nationwide and the planning and construction of new high-speed rail towns. Firstly, the planning and construction of new high-speed rail towns need to consider the negative impacts of existing stations and airports in the same city, as well as the distance of the station to major cities. Secondly, cities with the capability can enhance their high-speed rail benefits by developing the service sector. Thirdly, the spillover effect of high-speed rail stations is influenced by urban population size, but the relationship is not linear, so high-speed rail new town planning should consider local population size comprehensively.

The above case study aims to elucidate the role and application methods of the DID model. Regarding high-speed rail station spillover effects, many aspects still require further exploration. For instance, while light data are suitable for representing the scale or aggregation level of economic activities, they do not reflect the type of economic activities or industries. Therefore, research exploring the spillover effects of high-speed rail stations on a more detailed industrial basis is needed. Additionally, this paper concludes that city size and distance from major cities affect the spillover effects of high-speed rail stations, but the precise thresholds and mechanisms of influence remain to be further analyzed. Regarding the application of the DID model, it can be implemented using software like STATA, with detailed operations referenced in the STATA application manual. Further understanding of the DID model's principles can be gained from relevant books.

3.1.3 Application of Complex Network Analysis Method—Robustness Analysis of China's Railway Network

1. **Research Question**

As one of the most important infrastructures for economic development, railways have a lasting and profound impact on regional economic growth. They not only shorten travel time between cities but also affect the connectivity between cities, bringing many cities closer together. Regional economies benefit from reduced transportation costs, promoting economic growth along the lines (Jiao et al. 2016). In recent years, China's railway transportation has developed rapidly, with railways extending to all parts of the country, forming a network centered on railway lines and hubs.

With the high-speed and networked development of railways, the efficiency of connectivity between nodes in different locations has increased. However, this also creates potential crises, where the failure of a key station or line could lead to the dysfunction of other stations or lines, and in severe cases, paralyze a large part of the network. For instance, on September 8, 2017, due to signal failure, the Anhui section of the Beijing-Shanghai high-speed rail was suspended in both directions, affecting 46 train schedules and 6000 passengers. Natural disasters also often cause line collapses; for example, a 4.0 magnitude earthquake in Mianyang, Sichuan, on June 29, 2018, led to widespread train delays along the Sichuan-Chongqing high-speed rail, affecting many passengers' travel plans. China's railway passenger lines bear the main passenger transport task, and their safety and efficiency are crucial for national economic development and passenger safety. Therefore, the safe and stable operation of the railway network has become a research focus.

Most research on transportation networks is based on complex network theory, analyzing network topology and statistical features (Bagler 2008; Chen et al. 2020; Li and Cai 2004; Sen et al. 2003; Wang and Jin 2020; Wang et al. 2014, Zhang et al. 2010). However, key nodes affecting network operation are not necessarily those with high degree values. Consequently, network robustness research has emerged as a dynamic simulation method to better understand the topology (Chen et al. 2020). Network robustness refers to the ability of a network to maintain its original functions under random failures or malicious attacks. Albert is a pioneer in the reliability research of complex networks, measuring network damage after attacks using the largest cluster size, isolated clusters, and average path length (Albert et al. 2000). Holme observed network structural changes under different attack strategies and measured network performance using efficiency E and the largest connected subgraph S (Holme et al. 2002). Murray studied the robustness of the U.S. national road network by removing nodes, finding that damage to key sections could cause widespread paralysis of the network (Murray et al. 2007). Xu Feng constructed a high-speed rail-civil aviation composite network for topology and robustness analysis, finding that the network was less robust under deliberate attacks but more robust under random attacks (Xu et al. 2015b). Sun Xiaoxuan constructed three networks: the ordinary train subnetwork, the high-speed train subnetwork, and a dual-layer complex traffic network of high-speed and ordinary trains, finding that the reliability of the multi-layer traffic complex network was between that of high-speed rail and ordinary rail networks (Sun et al. 2019).

These studies show that when a network exhibits scale-free characteristics, it has key nodes whose attack can significantly impact network operation. Therefore, evaluating and measuring the importance of nodes in the network is crucial for enhancing system robustness and reliability and designing efficient system structures (Liu et al. 2006; Zhou et al. 2012). Current strategies for evaluating the robustness of transportation networks are basically consistent, i.e., examining changes in related metrics as the number of attacked nodes increases (Feng and Lei 2013; Jiang 2011). However, there are still some issues: firstly, traditional metrics for measuring network node importance (such as degree) have limitations in real-world operation. Railway networks are spatial entities existing in Euclidean space (Barthélemy 2011), and node importance evaluation should also consider the geographical characteristics of nodes, such as railway station location and distance between stations. Secondly, research on the robustness of transportation networks has focused on aviation networks (Chen et al. 2020; Sallan et al. 2014), urban road networks (Duan and Lu 2013; Lu et al. 2018), and subway networks (Zhang et al. 2010), with railway network robustness research needing further development. Thirdly, networks are often constructed based on the order of cities along railway lines (Guidotti et al. 2017; Lei and Qian 2015; Sun et al. 2019), representing only the connectivity between adjacent cities and limited to physical topological properties, neglecting railway operational functions. Additionally, identifying the main trunk of the network, which carries the main passenger transport in the network, is a controversial topic. Previous studies have identified hub cities based on capacity, centrality, and importance (Chen et al. 2020; Peng et al. 2019). Analysis based on network robustness offers a new dynamic perspective (Chen et al. 2020).

Therefore, this paper constructs a railway facility network based on complex network theory and a railway service network incorporating dynamic railway operation parameters (train operation paths, stop schedules, service frequencies, etc.). Combining the topological structure of the facility network with operational functions, the paper conducts quantitative measurements of network topological characteristics and simulations of network robustness, exploring the robustness of China's railway transportation network, and identifying main trunk cities in the national railway network, hoping to provide references for China's railway construction, station placement, and emergency response mechanism construction.

2. **Data**

The data used in this paper comes from the train schedule data published by the China State Railway Group Co., Ltd. in October 2020, which includes 59,164 passenger train services and 2822 railway stations across China (excluding Hong Kong, Macau, and Taiwan regions). The lines and stations were vectorized to produce a spatial distribution map of China's passenger railway network, as shown in Fig. 3.4. The railway network is unevenly distributed across the country. On the right side of the Hu Huanyong Line (Hu 1935), there is a dense network of lines and numerous stations, while on the left side, the network is sparse, with only a few main trunk railways. This distribution indicates that the infrastructure aligns with regional economic development levels, population concentration, and transportation demand.

3. **Complex Network Analysis Methods**

The network structure determines its function, and the network's ability to maintain integrity when nodes are under attack largely depends on the topological characteristics of the network (Kasthurirathna et al. 2013). We first adopt the research paradigm of complex networks to evaluate the topological structures of the railway facility network and the railway service network, respectively, and based on this, further simulate and analyze the robustness of the Chinese railway network.

A complex network is a research paradigm that represents complex systems as network structures (He 2009). By abstracting individuals as network nodes and the relationships between individuals as edges between nodes, a large number of real complex systems can be described as a complex network. There are two common modeling methods in complex network research: Space L and Space P. In the Space L model, only adjacent nodes on a route passing through multiple points are considered to be directly connected; in Space P, all nodes connected by a route are considered to be reachable by each other (Crespo and Garcia-Molina 2000).

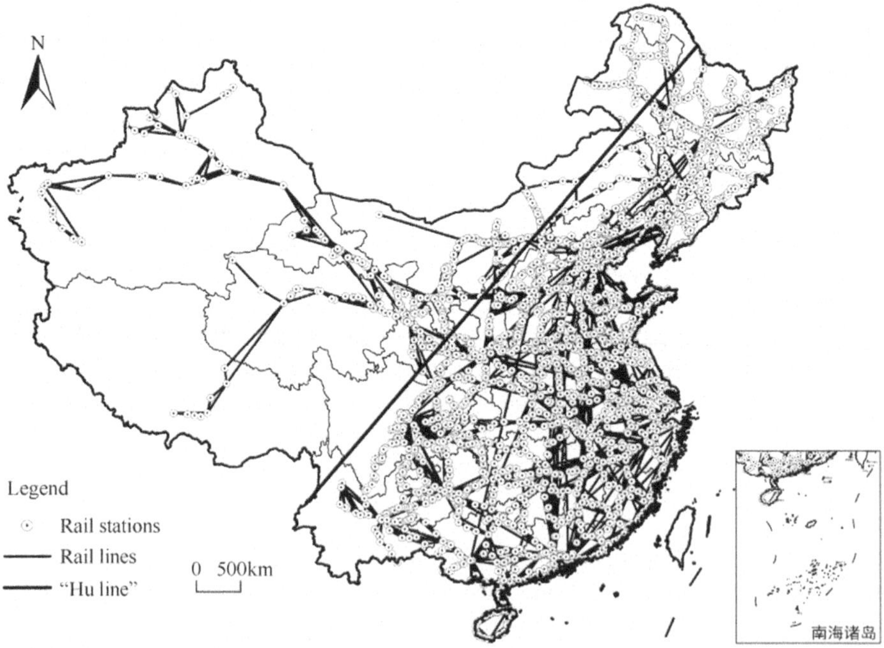

Fig. 3.4 China's railway transport network. *Note* Data for Taiwan unavailable

This study constructs the railway facility network based on Space L and the railway service network based on Space P. The network constructions under these two rules are shown in Fig. 3.5. Among them, Fig. 3.5a is a schematic diagram of the operation of trains D23, G6714, and G218, and Fig. 3.5b is the railway facility network established based on the tracks; by overlaying passenger information on Fig. 3.5b, c, which is the railway service network established based on the train operation scheme, is obtained. The railway service network shown in Fig. 3.5c can be used to evaluate the structure of the railway transportation network, but it is not suitable for evaluating the robustness of the railway transportation network since nodes on the route cannot be reached if a node is attacked and paralyzed.

By aggregating the stations in the same city, i.e., projecting the connection relationships between the original data stations to the city scale, the Chinese railway network $G = (V, E)$ is obtained. Here, the set of points V is the collection of cities, and E is the set of edges. This results in a railway facility network and a railway service network that include 318 cities (prefecture-level cities, prefectures, regions, municipalities) with 2252 edges in the facility network and 35,631 edges in the service network.

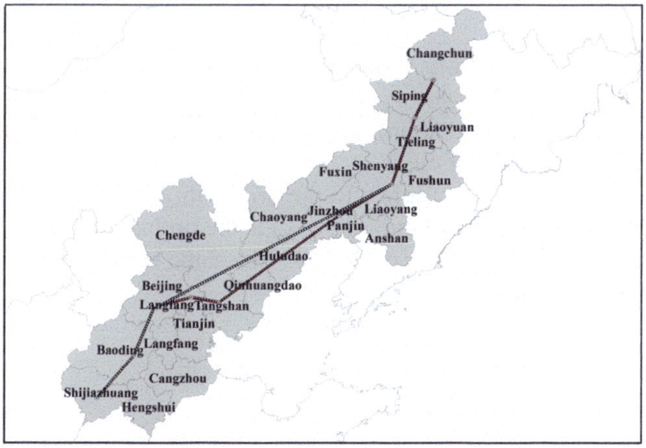

a. Operating schematic diagram of D23, G6714, G218

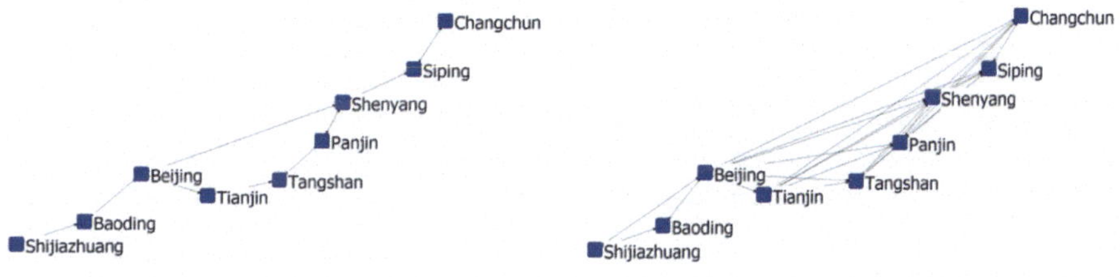

b. Railway physical network (RPN) c. Train service network (TSN)

Fig. 3.5 Network constructions of two modeling approaches

4. **Topological Characteristic Parameters of Complex Networks**

Degree Centrality (DC), Closeness Centrality (CC), and Betweenness Centrality (BC) are traditional indicators for evaluating the importance of network nodes.

(1) **Degree Centrality and Its Distribution**

Degree centrality is often used to evaluate the possibility and strength of a city's connections with other cities in the network. In unweighted networks, it is usually represented by the number of other nodes a node is connected to (DC), reflecting the city's connectivity; in weighted networks, it is often represented by the node's strength value (Strength Centrality, SC), reflecting the strength of the city's connections with other cities in the network. The calculation formula is shown as Formula (3.4).

$$\begin{aligned} DC_i &= \sum_{j}^{n} a_{ij} \\ SC_i &= \sum_{j}^{n} w_{ij} \end{aligned} \tag{3.4}$$

where a_{ij} refers whether there is a direct train connection between cities (if there is, this is 1; otherwise, this is 0), w_{ij} refers the frequency of train contact between cities (number of trains in operation), DC_i refers the node degree of city i, and SC_i represents the node strength of city i.

The degree distribution $P(k)$ refers to the probability that the node degree is k, and the cumulative degree distribution $P(> k)$ is the probability of occurrence of a node with a degree greater than or equal to k. (the same is true for the strength and the cumulative strength distributions).

$$P(k) = \frac{n_k}{n} \tag{3.5}$$

$$P(> k) = \sum_{k'=k}^{\infty} P(k') \tag{3.6}$$

(2) **Closeness Centrality**

The closeness centrality (CC) measures the extent to which a node is close to all other nodes along the shortest path. The CC describes the relative accessibility of cities in a network and is expressed as:

$$CC_i = \frac{N - 1}{\sum_{j \neq i} L_{ij}} \tag{3.7}$$

where L_{ij} is the length of the shortest path and N is the number of nodes in the network.

(3) **Betweenness Centrality**

The betweenness centrality (BC) measures the extent to which a particular node lies between other nodes in a network. The BC of a city reflects the transfer and connecting capacity of the network and is expressed as:

$$BC_i = \sum_{s \neq i \neq t} \frac{\sigma_{st}(i)}{\sigma_{st}} \tag{3.8}$$

where σ_{st} is the number of shortest paths between cities s and t and $\sigma_{st}(i)$ is the number of shortest paths between cities s and t via city i.

(4) **Shortest Path and Clustering Coefficient**

The shortest path between two nodes in a network is a path between two nodes that experiences fewer edges (Dorogovtsev and Mendes 2004). The shortest path L in the entire network refers to the average of the shortest paths between all pairs of nodes as:

$$L = \sum_{i=1}^{n} \sum_{i=j}^{n} \frac{L_{ij}}{2N(N-1)} \tag{3.9}$$

The clustering coefficient (CE) reflects the degree of aggregation between nodes in the network (Newman 2003), which is expressed as the ratio of the actual number of connections between neighbors of a node and the maximum number of connections between them as:

$$CE_i = \frac{|\{e_{st} : s, t \in N_i\}|}{\frac{k_i(k_i-1)}{2}} \tag{3.10}$$

where N_i is the point set containing all neighbors of node i; k_i is the number of elements in N_i (the number of neighbors of node i); s and t are the neighbors of node i; e_{st} is the number of connections between the neighbors of node i; and $\frac{k_i(k_i-1)}{2}$ is the maximum number of possible connections between neighbors. A large CE_i means that node i has a more compact system of connections with its neighbors (Hossain et al. 2013). In a fully connected network, the CE of all nodes is 1. Also, if node i has only one neighboring node, which means node i has a node degree of 1, then its CE_i is 0. The C of the overall network is the average of all individual CE, it values as:

$$C = \frac{1}{N} \sum_{v \in V} CE_i \tag{3.11}$$

5. **Network Robustness Evaluation Methods**

The evaluation of network robustness aims to dynamically simulate changes in a network during continuous attacks or node failures (errors, natural disasters, etc.). The relative size of the largest connected subgraph S and the global efficiency E are effective indicators to evaluate the network robustness. S describes the degree of fragmentation of the network and E evaluates global changes in the network structure (Chen et al. 2020).

(1) **Relative Size of Largest Connected Subgraph (S)**

The relative size of the largest connected subgraph (S.) refers to the ratio of the number of nodes in the largest connected subgraph to the number of all nodes in the network as:

$$S = \frac{N'}{N} \tag{3.12}$$

where N'N′. represents the number of nodes in the largest connected subgraph after the network is attacked, and N. represents the total number of nodes when it is not attacked (the original network). After some nodes have been attacked, the connectivity graph splits into countless small connectivity subgraphs. When there are only isolated nodes, S approaches zero.

(2) **Global Efficiency (*E*)**

The global efficiency (E) is defined as the reciprocal of the shortest distance between nodes:

$$E = \frac{2}{N(N-1)} \sum_{i>j} \frac{1}{d_{ij}} \tag{3.13}$$

where N is the total number of nodes in the network; d_{ij}. is the distance between nodes i and j. When i a j are not connected, $d_{ij} = +\infty$ and $E = 0$.

(3) **Network Attack Simulation Algorithm**

This study measures the robustness of RPN under different attack strategies. Figure 3.6 shows the flowchart of the network attack simulation. Random attacks simulate attacking nodes in thtwork at random: in each instance, a node is randomly selected and removed from the network, along with its edges. Intentional attacks prioritize attacking nodes that occupy important positions in the network, rendering the connected edges ineffective. If multiple nodes have the same attribute values, one is randomly removed to ensure that only one node is attacked at a time, until all nodes in the network are removed. Based on the two networks constructed in this study (RPN and TSN), intentional attacks are divided into two modes. In intentional attack mode 1, the network is attacked in the order of node importance in TSN, based on the initial state of the network without re-evaluating the importance of nodes; in intentional attack mode 2, the order of attack is based on the evaluation results of node importance in RPN. After each attack, the most important node in the remaining network is determined based on the newly generated adjacency matrix, and this node and its connected edges are attacked until the network is paralyzed.

6. **Result Analysis**

(1) **Structure of China's Railway Transportation Network**

(i) **Railway Network Degree and Its Distribution**

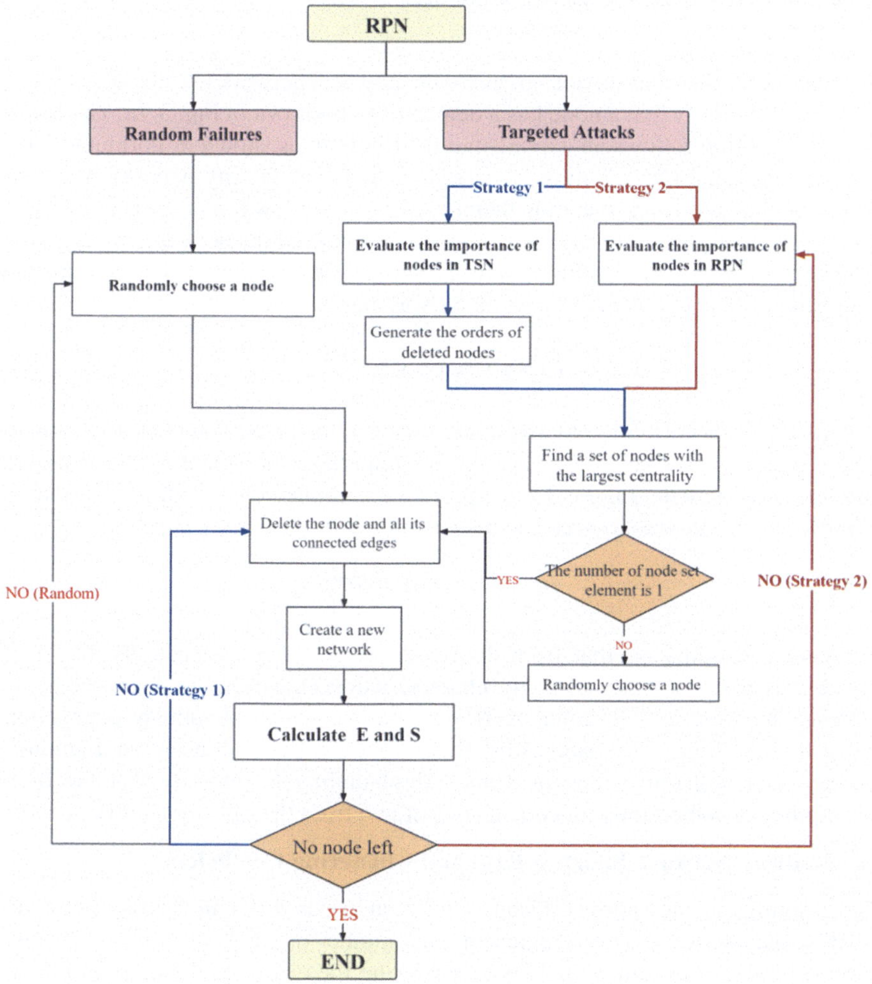

Fig. 3.6 Railway network robustness evaluation flowchart

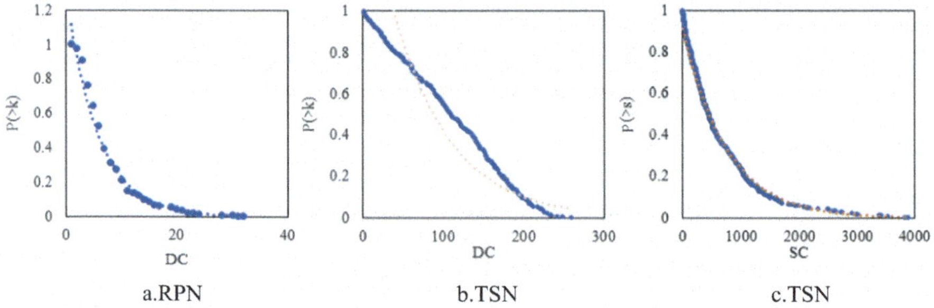

Fig. 3.7 Network's DC and SC cumulative distribution

The cumulative distribution of degrees for the physical network (RPN), which represents the probability that a node has a degree of k, is shown in Fig. 3.7a. The cumulative distribution follows an exponential distribution, as shown in Formula (3.14). This indicates that China's railway passenger transport facility network possesses scale-free characteristics, meaning that most cities have low direct connectivity with other cities through the railway, positioned at the edges of the network. A very small number of stations act as traffic hubs in the network, having a high degree value and connecting different routes to maintain railway operations.

$$P(>k) = 1.33e^{-0.176k} \tag{3.14}$$

Figure 3.7b shows the cumulative degree distribution of the Train Service Network (TSN), which poorly fits an exponential function, thus not validating the scale-free network characteristic. Figure 3.7c displays the cumulative strength (s) distribution of TSN, which follows a power-law distribution, as indicated by Formula (3.15).

$$P(>s) = 0.91e^{-0.001s} \tag{3.15}$$

Figure 3.7b indicates that from the perspective of city connectivity, the intercity railway network lacks scale-free characteristics. However, the scale-free nature becomes apparent in TSN after overlaying the frequency of railway connections between cities (Fig. 3.7c), suggesting that major routes are concentrated among a few city pairs, with only a few important nodes handling the majority of connections. This finding is consistent with conclusions drawn from the railway facility network.

(ii) **Railway Network Shortest Path and Clustering Coefficient**

The meanings of the shortest distance between nodes differ in the two networks. In RPN, the shortest distance represents the number of edges (or nodes) one must traverse between two cities; in TSN, considering the operational factors of railway routes, the shortest distance signifies the number of line transfers needed between two cities. $L_{ij}=1$ indicates that two cities can be reached directly by one train without transfers, while $L_{ij}=2$ indicates that one transfer is needed. According to Formula

3.9, RPN's L = 4.06, and TSN's L = 1.72, showing that the average distance between city nodes in the facility network is 4, meaning an average of four stops to reach a destination, while the service network indicates that passengers need on average only one transfer to reach their destination.

Using Formula 3.11, it is calculated that the average clustering coefficient for the railway facility network is 0.52, and for the service network, it is 0.73. TSN's clustering degree is significantly higher than RPN's, indicating a very low probability of global coupling between adjacent nodes in the facility network, leading to certain sections becoming inoperable if a node fails. Moreover, both TSN and RPN have an average clustering coefficient much higher than that of a random network of the same scale (0.045), and close to the clustering coefficient of small-world networks (0.635), suggesting that both networks are typical small-world networks, meaning most nodes are not directly adjacent but can be reached through relatively short paths.

Furthermore, the study found a significant negative correlation between the clustering coefficient and degree centrality in TSN, as shown in Fig. 3.8. Nodes with higher degree values have fewer direct connections between their neighboring nodes, indicating weaker inter-city connections around them, forming a hub-and-spoke structure. Additionally, Fig. 3.7b shows TSN lacks scale-free characteristics, proving that the railway network, unlike the aviation network's scale-free operational system (Wang and Jing 2017), presents a more complex hierarchical structure.

(2) **Railway Transportation Network Robustness**

(i) **Railway Network Robustness Simulation**

Based on the algorithm described previously, we developed a railway transportation network robustness simulation evaluation algorithm to assess the robustness of the railway network under two attack strategies: random and intentional attacks. Since

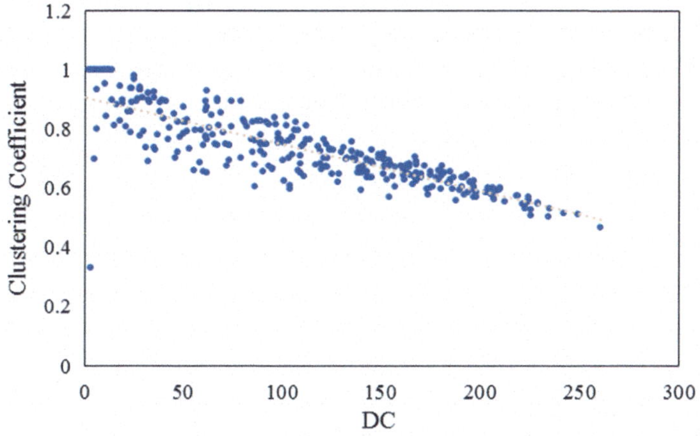

Fig. 3.8 Correlation between degree centrality and clustering coefficient in the railway service network

the robustness of the railway network depends on the facility network, with trains only operational if the facility network is unobstructed, all robustness evaluations are conducted based on the facility network (RPN). RPN's initial global network efficiency (E) is 0.27, and the initial size of the largest connected subgraph (S) is 1.0.

Random Attack

The random attack model aims to analyze the change in network robustness when nodes are subject to random attacks (e.g., failures, natural disasters), where all cities have an equal probability of being attacked. Figure 3.9 shows the changes in S and E of the railway facility network (RPN) during a random attack. It is observed that in the first 100 attacks, nodes are removed one by one, causing a nearly linear decrease in S, but a relatively slow decline in E. When about half of the nodes are removed, both E and S experience a rapid decline, with the global efficiency halving, indicating that RPN relies on approximately 50% of its nodes to withstand random attacks. Additionally, RPN only collapses completely near the end of the attacks, that is, S and E tend toward 0 (a brief increase in E before the attacks conclude might be due to remaining nodes locally networking and interconnecting), showing that RPN is relatively robust under random attack scenarios.

Intentional Attack

Intentional attacks target nodes based on their importance, simulating the robustness of the railway network under deliberate attack scenarios. The order of attack is determined by the importance of network nodes, with significant differences between evaluations of node importance in RPN and TSN. From the RPN perspective, the importance of node cities reflects their topological attributes within the network, whereas from the TSN perspective, it reflects the city's convenience within the railway transportation network, indicating its importance in the national socio-economic system. Consequently, the latter holds more practical significance and is more likely to be

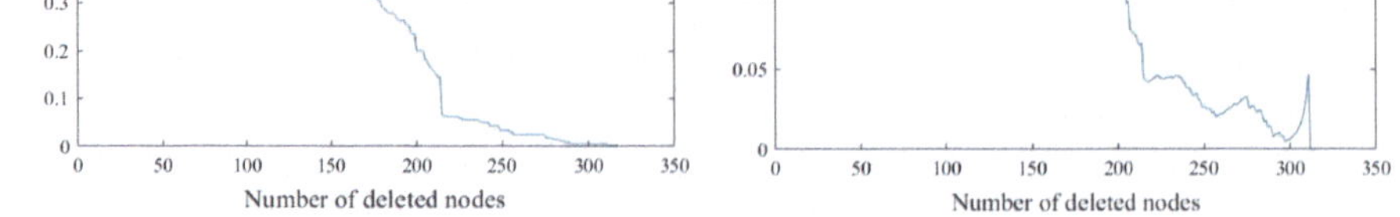

a. Changes in *S* b. Changes in *E*

Fig. 3.9 Changes in S and E of RPN under random attack

targeted in scenarios like wars or terrorist attacks. Thus, we simulated the robustness of the railway network under two attack modes: intentional attack mode 1, based on the importance of node cities from the TSN perspective, without re-evaluating node importance during the attack process; and intentional attack mode 2, based on the importance of node cities from the RPN perspective. During intentional attacks, RPN is continuously damaged, its topological structure changes, and the topological attributes of node cities evolve, necessitating the re-evaluation of importance indicators (DC, BC, CC) after each attack in mode 2. The evaluation results can inform the development of railway transportation emergency plans.

Figure 3.10 shows the changes in the largest connected subgraph (S) and global network efficiency (E) during intentional attack mode 1. Both S and E initially decrease rapidly, then slow down, with the turning point for the decreasing trend of S at about 100 nodes: that is, S decreases quickly before reaching 100 nodes. When approximately 100 nodes are removed, E drops to a lower level, then the rate of decline slows, indicating that the first 100 cities primarily carry the network's operational functions. Additionally, Fig. 3.10 shows that the attack results based on BC lead the decrease trend in the size of the connected subgraph (S) to approach zero first. Regarding changes in global network efficiency (E), the decline rate of the curve based on BC is the fastest, and network efficiency reaches zero first. The attack effects based on CC and DC are similar, with a high degree of overlap in the changes in global efficiency (E). The brief increase in E towards the end of attacks based on CC and DC can be explained by the possibility of locally networked, remaining low-importance nodes forming local connections. Overall, the attack strategy based on BC is the most effective in intentional attack mode 1, making it the weakest link in China's railway transportation under this mode.

Figure 3.11 shows the changes in S and E of RPN during intentional attack mode 2. As RPN continuously suffers damage and its topological structure changes, the topological attributes of node cities also evolve, requiring the re-evaluation of importance indicators (DC, BC, CC) after each attack in this mode. According to Fig. 3.11, the

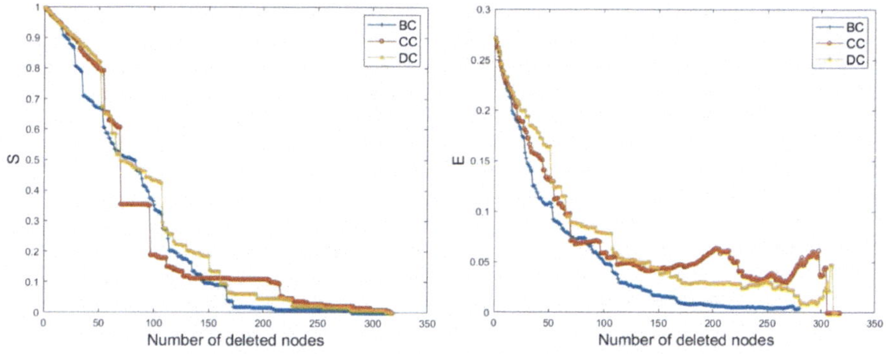

a. Changes in *S*

b. Changes in *E*

Fig. 3.10 Changes in S and E of RPN under intentional attack (mode 1) attack results based on DC, BC, and CC show strong consistency; S and E plummet from the first attack, with global network efficiency halving when about 10% of nodes are attacked; the network completely fails (S and E reduce to zero) when approximately 100 nodes are removed, splitting into countless subnetworks or isolated nodes. The steepness of the decline in the size of the largest connected subgraph (S) based on BC and CC is the steepest before reaching 20% of nodes removed, indicating that 20% is a critical threshold where the network maintains normal operations.

From this analysis, first, the railway facility network (RPN) is robust under random attacks but fragile under intentional attacks. Second, under intentional attack mode 1, the network collapses (E and S reduce to zero) around 300 nodes, while in mode 2, the critical point is around 100 nodes, indicating that RPN fails faster under dynamic intentional attack strategies based on the importance evaluation from the TSN perspective (mode 2), making it a more effective attack method. Lastly, the results of intentional attacks in both modes show that the attack strategy based on BC (mode 2) decreases the fastest, making it the most effective strategy against China's RPN. This is because BC, as an indicator of nodal hub status, sequentially removing cities with the strongest intermediacy (BC) causes the most extensive route failures.

(ii) **Railway Network Fragmentation Process—Robustness Spatial Analysis**

To better understand the impact of various attacks on the network structure, we further analyzed the morphological changes and robustness conditions of RPN during the network fragmentation process. The analysis shows that intentional attacks based on the betweenness centrality (BC) of nodes from the TSN perspective are the most effective strategy, leading us to conduct a spatial analysis of RPN's robustness based on this attack mode. We designate the removal of 5% of nodes as the first attack, 10% as the second, 20 and 30% as the third and fourth attacks, respectively. To analyze the railway network's fragmentation process, we examined the impact on provincial capitals and municipalities during intentional attacks, analyzing spatial

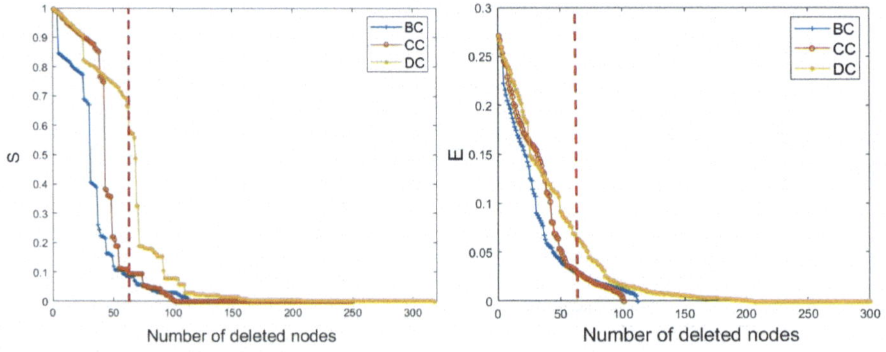

a. Changes in *S*

b. Changes in *E*

Fig. 3.11 Changes in S and E of RPN under intentional attack (mode 2) differentiation based on the four major regions: Northeast, East, Central, and West. The Northeast includes Liaoning, Jilin, and Heilongjiang provinces; the East includes Beijing, Tianjin, Shanghai, Hebei, Jiangsu, Zhejiang, Fujian, Shandong, Guangdong, and Hainan; the Central includes Shanxi, Anhui, Jiangxi, Henan, Hubei, and Hunan; the West includes Chongqing, Guangxi, Sichuan, Guizhou, Shaanxi, Gansu, Qinghai, Tibet, Inner Mongolia Autonomous Region, and Ningxia Hui Autonomous Region.

Table 3.8 calculates the percentage of provincial capitals that failed in each attack relative to the total number of new failures, and similarly for nodes located in the Northeast, East, Central, and West regions. The results show that the proportion of provincial capitals failing in the first attack reached 63%, and all were removed by the fourth attack, highlighting the importance of provincial capitals to the network's overall robustness. Among the four attacks, the proportion of cities removed in each region was East (35%) = Central (35%) > West (25%) > Northeast (19%); within each attack, the East had the highest proportion in the first (7%) and third (16%) attacks, while the Central region had the highest in the second (10%) and fourth (11%) attacks. Based on the strategy of removing important nodes first, it is believed that the East and Central regions play a decisive role in China's railway network structure; the proportion of Western cities removed was relatively even across the four attacks, indicating their moderate importance; the Northeastern nodes concentrated in the third and fourth attacks, suggesting their lower importance from a robustness perspective.

During each attack, the fragmentation process of RPN is shown in Fig. 3.12 (red nodes indicate the first attack targets, blue for the second). As shown in Fig. 3.12a, in the first attack, the failure of Beijing, Tianjin, and Shijiazhuang caused the Northeastern network to form a partial web, disconnecting from the main network; failures originating from Wuhan and Zhengzhou significantly reduced the rail network density in the Central region, demonstrating their role as transportation hubs; in the West, the failures of Xi'an and Lanzhou in the first two attacks caused the already sparse Northwestern network to disconnect from the main trunk, severing connections with the Central and Eastern regions. When 20% of nodes were attacked, the railway network became significantly sparse (Fig. 3.12c), with less than 30% of the original operational routes remaining, mostly locally connected, especially within the Shandong Peninsula where city nodes maintained frequent connections, minimally

Table 3.8 Spatial distribution of failed cities during the four attacks

	Attack I (5%)	Attack II (10%)	Attack III (20%)	Attack IV (30%)	Total
Provincial capital cities (municipalities)	63%	19%	28%	13%	/
Eastern (%)	7	3	16	9	35
Central (%)	6	10	8	11	35
Western (%)	5	5	6	9	25
Northeastern (%)	0	0	11	8	19
Total number of nodes attacked	16	16	32	32	96

affected by node failures. At 30% node failure (Fig. 3.12d), only a few spatially proximate cities could connect, aligning with the conclusion that the network approached paralysis at 20% node removal (Fig. 3.11), reaffirming the rationality of a 20% threshold for cities maintaining the railway network's health and basic functions under intentional attack mode 2 based on BC.

(iii) **Network Backbone Analysis**

The network backbone consists of subnetworks that carry the main passenger traffic, determining the entire network's operational efficiency. If the backbone is attacked, the entire network can quickly collapse. Previous analysis indicated that intentional attacks based on BC of nodes in the facility network are the most effective way to paralyze the railway network. Therefore, we determined the network backbone based on the order in which cities were removed under two intentional attack strategies. Dynamic simulation results of railway network robustness showed that RPN's normal operation depends on 20% of the node cities, forming the backbone of China's railway passenger network based on the first 20% of nodes removed under both attack strategies.

Figure 3.13a and b display the main trunk of China's railway network under strategies 1 and 2 (based on TSN and RPN) and the train routes passing through the backbone. The former represents the backbone of China's Train Service Network (TSN), closely related to the distribution pattern of population and economy, hence, TSN's backbone reflects the economic connections between regions in China. The latter represents the backbone of China's railway facility network, crucial for supporting normal railway operations. Both backbones appear similar, but the main trunk cities differ significantly. Visually, the former stretches from the Northeast to the Southwest, while the latter is primarily concentrated in the Central region. In TSN's backbone, economically developed cities in the Central and Eastern regions occupy major positions, but in RPN's topological structure, nodes of higher importance are mainly distributed in the Central region, indicating that RPN's backbone in the Central region is more likely to lead to network paralysis.

Moreover, Fig. 3.13b shows the main network trunk and the train routes passing through it, clearly revealing the North–South Beijing-Guangzhou and Beijing-Shanghai lines, the Northeast's Beijing-Harbin line, and the East–West Shanghai-Wuhan-Chengdu, Shanghai-Kunming high-speed, and Beijing-Lanzhou lines forming the network's backbone. Beijing, Tianjin, Shijiazhuang, and Langfang serve as major passenger distribution centers in the North China region, connecting major routes to the Northeast; Central cities like Wuhan and Zhengzhou have a significant impact on the overall network; Xi'an and Ulanqab are key cities for routes heading Northwest and towards Inner Mongolia. From the perspective of network robustness, identifying these key node cities is crucial for effectively protecting them during passenger line transportation, ensuring the overall stability and efficiency of the transport network.

Given that intentional attack strategy 2 is the most effective way to paralyze China's railway transportation network, further analysis was conducted on the main trunk cities identified under this attack mode. Table 3.9 lists the nodes targeted in

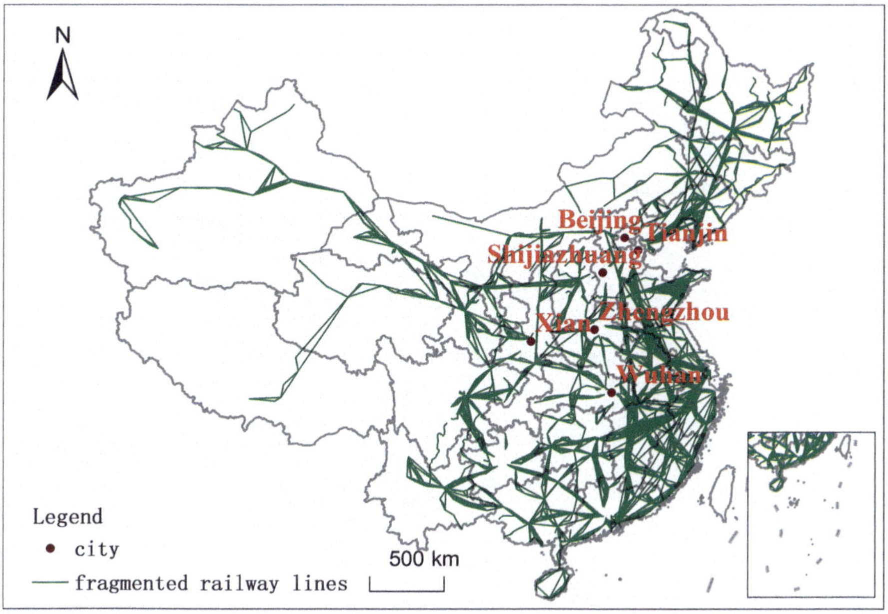

(a) 5% nodes deleted

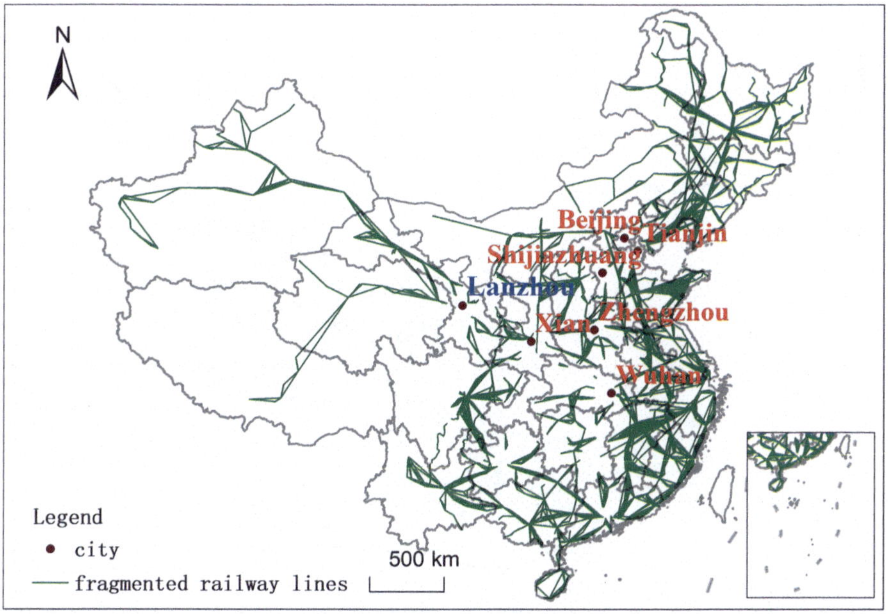

(b) 10% nodes deleted

Fig. 3.12 Railway network fragmentation process based on BC in RPN perspective. *Note* Data for Taiwan unavailable

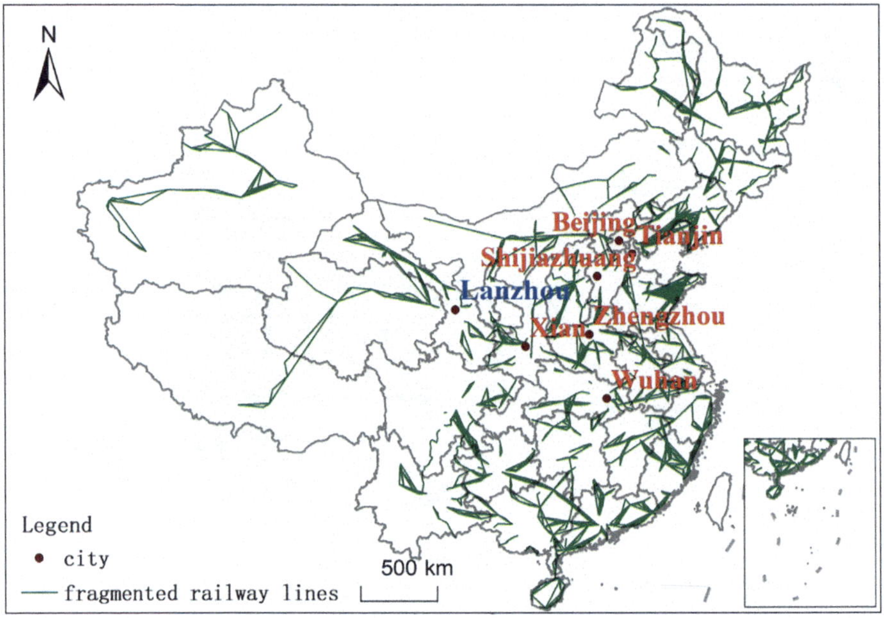

(c) 20% nodes deleted

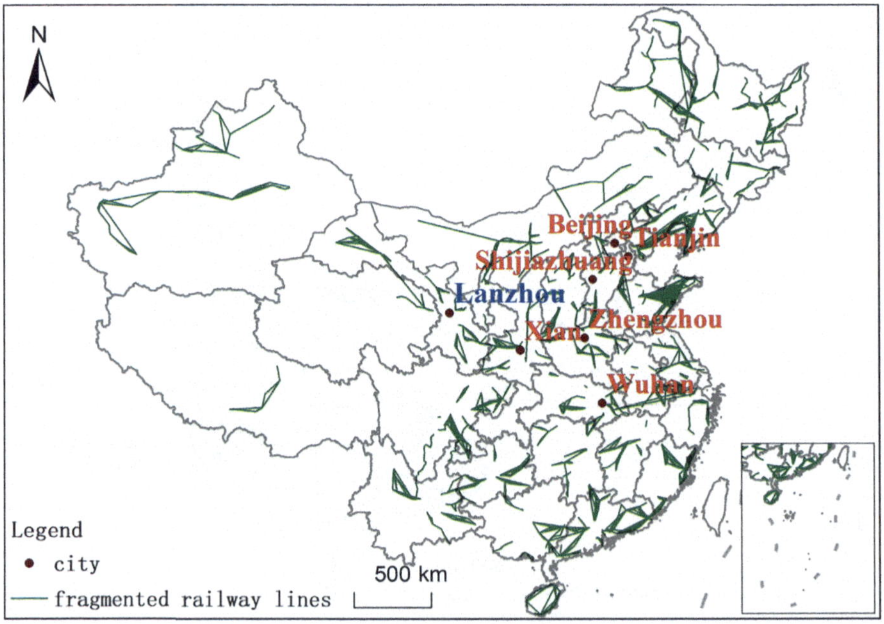

(d) 30% nodes deleted

Fig. 3.12 (continued)

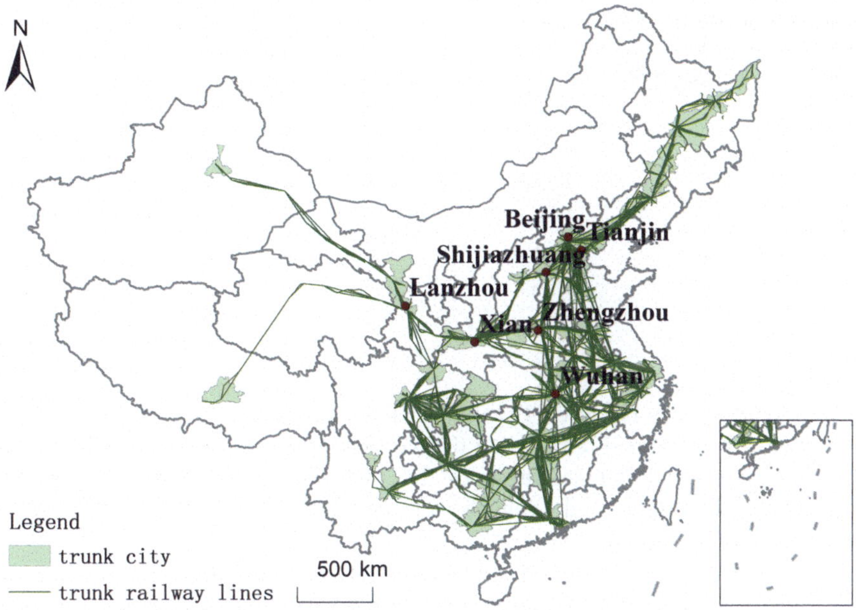

(a) Intentional attach strategy 1 (based on TSN)

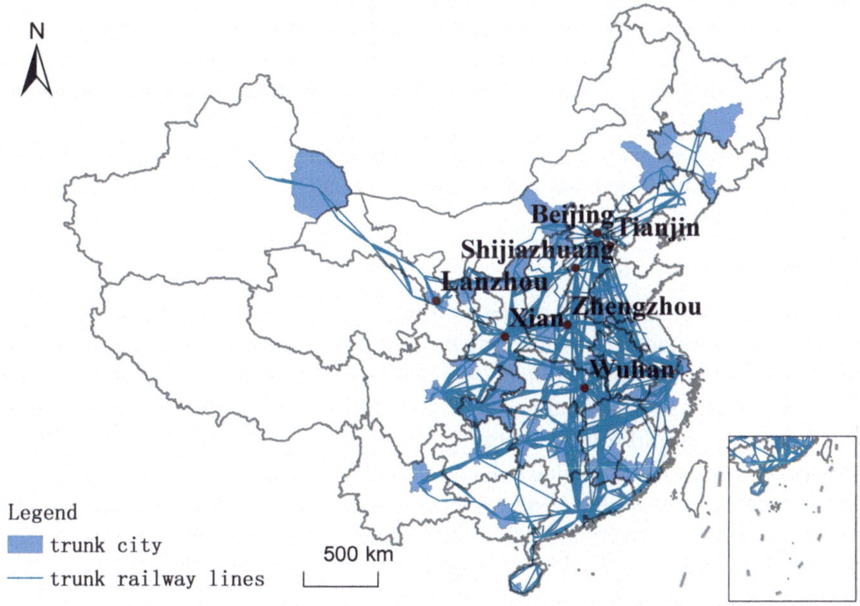

(b) Intentional attach strategy 2 (based on RPN)

Fig. 3.13 Identification results of China's railway passenger network backbone. *Note* Data for Taiwan unavailable

Table 3.9 Comparison of the top eight nodes for BC attacks (Strategy 2) and their original BC rankings

City	BC attack order	BC ranking
Beijing	1	1
Tianjin	2	40
Shijiazhuang	3	11
Langfang	4	165
Ulanqab	5	149
Wuhan	6	4
Xi'an	7	2
Zhengzhou	8	6

the first eight attacks based on BC under RPN evaluation, along with their initial BC ranking. The order of attack and degree ranking show significant differences, indicating substantial changes in nodes' BC after each attack. Cities like Tianjin, Shijiazhuang, Langfang, and Ulanqab were targeted earlier, while Wuhan, Xi'an, and Zhengzhou were delayed.

After Beijing was removed, Tianjin, originally ranked 40th in BC, became the node with the strongest transfer attribute in the network, suggesting that many routes originating from or passing through Beijing could detour through Tianjin in case of a fault in Beijing. Similarly, Shijiazhuang, Langfang, and Ulanqab, geographically close to Beijing, remain "silent" under normal operations; their significance emerges if Beijing encounters a fault, helping to divert some passenger flow and maintain network operations to a certain extent. Conversely, Wuhan, Xi'an, and Zhengzhou, initially high in BC ranking, were delayed, indicating that some high-BC nodes are not as crucial to the overall network, highlighting the difference between dynamic robustness simulation and static topological indicators like K-core for identifying main trunk cities.

7. **Conclusion**

As China's railway network continues to improve, railway transportation plays an increasingly important role in ensuring the mobility of the population and promoting stable socio-economic development. A stable and reliable railway network is crucial for the daily travel and material transportation needs of the country's residents. Accidents leading to partial route disruptions and temporary station closures in China's high-speed railway network, due to internal faults or force majeure, affect the service capability of the network to varying degrees. Understanding the structure and robustness of the railway network is vital for drafting emergency plans, enhancing the efficiency of the railway network's transportation, and ensuring its balanced operation. This paper analyzed the structure of China's railway network using complex network analysis algorithms, simulated its robustness, and identified the backbone of China's railway transportation network. The specific conclusions are as follows:

(1) Railway complex networks constructed under different rules exhibit distinct topological properties. The Railway Facility Network (RPN) is a typical scale-free small-world network, while the Railway Service Network (TSN) shows a negative correlation between degree, degree distribution, and clustering coefficient. Unlike aviation, railway operation strategies contribute to a complex networked or hierarchical structure in the railway service network.
(2) Network robustness simulations indicate that the railway network is robust against random attacks but fragile under intentional attacks. The critical number of nodes for network collapse (where E and S reduce to zero) from the perspective of TSN and RPN is approximately 300 and 100, respectively, suggesting that RPN fails faster under dynamic intentional attack strategies based on node importance evaluation. Intentional attacks based on the BC value of nodes are the most effective attack strategy, with a consistent conclusion in both modes: a 20% threshold, where the rapid decrease in robustness indicators occurs upon removing the first 20% of nodes, beyond which the network is relatively stable. Thus, 20% of the cities bear the main railway network's operational routes.
(3) Dynamic intentional attacks based on BC from the RPN perspective identified a backbone network consisting of the 20% most significant parts of the entire network, whose collapse would paralyze the railway network. The identification process of the railway network backbone revealed the importance ranking of cities, with key nodes being Beijing, Tianjin, Shijiazhuang, Langfang, Ulanqab, Wuhan, Xi'an, Zhengzhou, Guangzhou, Changsha, Fuyang, etc. The spatial distribution of main trunk cities clearly forms the main framework of the North–South Beijing-Guangzhou and Beijing-Shanghai lines, the Northeast's Beijing-Harbin line, and the East–West Shanghai-Wuhan-Chengdu, Shanghai-Kunming high-speed, and Beijing-Lanzhou lines.

Future research needs to delve deeper into several areas: Firstly, due to the complexity of the railway network, its robustness represented only by global efficiency (E) and the largest connected subgraph (S) appears inadequate. Further evaluation metrics, such as relative entropy, should be considered. Secondly, the functional characteristics of the railway network affect the results of robustness evaluation, necessitating the establishment of a directed vector network and further evolutionary studies considering train frequency, travel time, and passenger flow. Additionally, when a railway station fails, nearby cities become alternative destinations; cities with more alternative destinations might see their importance decrease. Therefore, it is necessary to consider both topological properties and spatial proximity when evaluating node importance.

3.2 Urban Spatial Simulation Model

3.2.1 Overview of Urban Spatial Simulation Models

Urban spatial simulation models are further deductions of urban spatial development laws and applications of urban space evolution rules. These models simulate the urban spatial development process, predict future conditions, and assess policy impacts. Since the late 1960s, computer-assisted urban simulation technology has been introduced into urban planning. The academic research on spatial economic connections shifted from simple qualitative, static descriptions to more dynamic decomposition and the application of quantitative models, with urban space evolution process simulation becoming a focus of attention, leading to the development of influential models. Professor Batty categorized them into two types: models based on Cellular Automata (CA) and Agent simulation technology, and the Lowry model series (Batty 2008; Batty et al. 2013).

Models based on CA and Agent simulation technology offer a bottom-up, micro-to-macro simulation approach (Batty 2008, Batty et al. 2013). Specifically, they study the macroscopic dynamic evolution of urban space by simulating the behaviors of a large number of microscopic agents and their interactions, providing a better grasp of uncertainties in urban systems (Ahmed et al. 2017; Dai et al. 2019; Krzysztof 2017; Liu et al. 2018; Long et al. 2014; Stanislaw and Yang et al. 2016), and have been applied in urban spatial simulations (Chao et al. 2007; Li et al. 2013; Shen et al. 2009; Yu et al. 2012). Current Agent-based studies are expanding from land use changes to population and social domains (Tian 2019; Qiu et al. 2019). However, CA simulation technology typically requires detailed individual behavior data at the micro-level, which is greatly limited by data availability. Due to the large behavioral differences among individuals (e.g., different market behaviors of companies of varying sizes), describing individual behaviors is extremely complex, posing a major modeling challenge (Batty et al. 2013). Moreover, during the simulation, agents evolve randomly, leading to variations in each run's results, affecting the accuracy and reliability of conclusions.

The Lowry model series refers to a category of LUTI models developed and derived from the traditional Lowry model framework. These models usually process and output aggregated data sets, converging various driving forces to an equilibrium state, representing a top-down simulation process (Batty et al. 2013). These models excel in simulating the macroscopic pattern of urban socio-economic activities. Lowry established the first LUTI model in 1964 to simulate urban spatial development trends, known as the Lowry model (Lowry 1964). The Lowry model divides the urban economic sector into basic and service sectors, specifying the location of the basic sector and outputting the spatial distribution of residents and service sectors. Subsequently, many scholars have conducted empirical analyses and further developed the Lowry model (Wang 1998). In recent years, due to the continuous development of computer technology and the increasingly significant trend of interdisciplinary integration, more and more new methods and theories have been introduced into LUTI model research, developing a series of models such as TOPAZ, DRAM-EMPAL, OSAKA, DELTA, MEPLAN, etc. (Aljoufie 2014; Brandi et al. 2014; Coppola et al. 2013; Echenique et al. 1990; Niu et al. 2019; Simmonds and Feldman 2011; Zondag et al. 2015). These works have played an important supporting role in assisting urban spatial decision-making, gradually reflecting the dynamic and complex characteristics of urban space, marking a significant breakthrough in urban model development.

3.2.2 CA Model

Since the 1990s, with the maturation of GIS technology and its widespread application in urban spatial simulation research, coupled with advances in complexity science and artificial intelligence technology, new concepts and algorithms have been brought to urban simulation models, with CA being a typical example. CA models are grid-based dynamical models with discrete time, space, and states, where spatial interactions and temporal causality are local. Their "bottom-up" research approach, powerful complex computational capability, inherent parallel computing capacity, high dynamism, and spatial concept make them highly capable of simulating the spatiotemporal evolution of complex spatial systems, offering a natural advantage in geographical research (Zhou et al. 1999). Additionally, as naturally integrated spatiotemporal models with a discretely divided spatial structure, CA models excel in simulating geographic complex systems with spatiotemporal characteristics (Chen et al. 2000). CA models lack explicit equation forms but include a series of model construction rules, focusing on spatial interactions rather than socio-economic indicators' influence, better reflecting spatial pattern changes and further feedback effects. Moreover, because the model's cellular space can be very finely divided, it can finely represent changes in urban spatial structure; additionally, the model can reflect the city's lifecycle from emergence, development to demise over a longer time scale. CA has been earliest and most deeply researched in simulating urban growth, diffusion, and land use evolution, and remains a hot topic in CA applications (Chen 1999b).

1. **Basic Principles of CA Model**

CA was first proposed by the renowned mathematician John von Neumann in the 1940s as a model based on interactions between microscopic units, evolving to simulate complex systems, distinct from traditional linear equation-based models. CA reflects complex system characteristics such as emergence, ambiguity, and evolution (Chen et al. 2002; Wolfram 1984). Unlike system dynamics models, CA models use a "bottom-up" approach to simulate complex systems, composed of rules rather than strict physical or mathematical equations, thus serving as a methodological framework and a general term for a class of models (Chen et al. 2002).

The Cellular Automata (CA) model is composed of cells, cell states, neighborhoods, and transition rules. A cell is the basic unit of a CA model, and each cell can have multiple state variables, but at any given moment, it only has one state, which is taken from a finite set (this set could be a binary set like "0" or "1", or a discrete set containing various variables). A neighborhood is a collection of cells determined by a specific rule around a central cell, which affects the state of the central cell at the next moment. There are typically four types of neighborhoods, such as the von Neumann neighborhood and the Margolus neighborhood. The transition rule is a function that determines how the cell state changes over time based on the neighborhood, and it is the core part of the CA, determining the final outcome of the simulation. This can be represented by the following formula (Amormso and Patt 1972):

$$A = (d, s, N, f) \tag{3.16}$$

In the formula, A represents the CA system, d denotes the dimensionality, s stands for the cell state, N indicates the neighborhood, and f is the local transition function.

Analyzing the structure and rules of cellular automata, a standard CA model should exhibit the following characteristics (Cao 2010; Zhou and Li 2012):

(1) Homogeneity and Uniformity: Homogeneity means that the changes in every cell within the cellular space follow the same rules, which are the rules of the cellular automaton model. Uniformity refers to each cell having the same size, shape, and distribution, with a regular spatial distribution pattern.
(2) Spatial Discreteness: Each cell is discretely distributed on a grid point according to specific rules, and its state evolves based on deterministic rules.
(3) Temporal Discreteness: The state of a cell at one moment only affects the state of the cell in the next moment.
(4) Discrete and Finite States: Each cell can only have a finite number of states. Dynamic systems with continuous states often do not need to undergo granulation to be converted into symbolic sequences. In practice, some continuous variables often need to be discretized, such as through classification or grading, to facilitate the construction of cellular automaton models.
(5) Synchronous Computing (Parallelism): Changes in the state values of each cell are independent of each other and do not influence one another. If the process of changes in the cellular automaton's configuration is viewed as data or information processing or computation, then the processing in cellular automata is synchronous, making it particularly suited for parallel computing.
(6) Spatio-Temporal Locality: The state of a central cell at a certain moment is only related to the states of cells in its neighborhood; from the perspective of information transmission, the speed of information transfer in cellular automata is often limited.
(7) High Dimensionality: With k states for each cell and n cells, theoretically, CA's evolution rules number k^{k^n}, endowing CA with a strong capability to model complex systems.

Urban spatial evolution generally manifests as the derivative expansion of existing urban land. From a microscopic perspective, it actually corresponds to changes in the land use status of different plots in a city. The change in the land use status of plots is influenced by the land use status of surrounding plots in reality, similar to how a cell's state in a CA model is influenced by the states of its neighbors. This similarity provides a basis for linking urban land expansion with cell growth in the CA model, suggesting an isomorphism between the mechanisms of urban spatial growth and evolution and the cell growth mechanism in the CA model. Specifically, each plot in a city can be viewed as a cell in the CA model, with the plot's land use status corresponding to the cell's state, and the agglomeration effect of urban evolution and development analogous to the neighborhood effect in the CA model. The rule for converting non-urban land into urban land in cities is similar to the state transition rules in CA. Clearly, using the CA model to simulate the process of urban spatial evolution is rational.

The advantages of CA in simulating complex systems have led to its widespread application in urban expansion simulation. Using the CA model offers the following benefits (Li 2007):

(1) CA can be easily coupled with GIS: In terms of data exchange, both CA and GIS models can conveniently process raster data. In data processing, both models use discrete methods for data processing, facilitating data exchange.
(2) The "bottom-up" simulation approach of CA aligns closely with the self-organizing phenomenon in urban land use expansion, making CA suitable for simulating urban land use expansion.
(3) The simulation process of CA models simultaneously covers time, space, and states, treating these three elements as equally important in the simulation, ensuring consistency in form and function of CA.
(4) CA is a dynamic model capable of simulating the dynamic evolutionary process of urban land use expansion, offering advantages over static models.

2. **Applications of CA Models in Urban Spatial Simulation**

CA models have significant advantages in simulating the dynamic evolution of complex systems. Since the 1990s, an increasing number of scholars have used CA models for economic growth and scenario analysis of urban systems, leading to many influential urban CA models.

In the 1970s, Tobler used a CA model to simulate urban expansion around the Great Lakes region of Detroit, USA, marking the first introduction of the concept of cellular automata into geographic research (Tobler 1970). In the 1980s, Couclelis extensively discussed the application of CA models in urban diffusion, arguing that the uncertainty of urban development policies and simulations determined the wide application of CA models. She also conducted in-depth research on the integration of CA models with Geographic Information Systems (GIS), significantly influencing the study of simulating urban diffusion with cellular automata (Couclelis 1985, 1989).

Entering the 1990s, more case applications of CA models emerged in urban spatial simulation research. For example, White and Engelen (1993) applied a cellular automaton model to simulate urban land use changes in Cincinnati, USA; Deadman et al. (1993) used a CA model to predict the distribution pattern and expansion process of rural settlements in Ontario, Canada, demonstrating good accuracy in the model results. Other scholars applied CA models to studies on urban morphology, urban expansion, and urban land use conditions (Cecchini 1996; Phipps 1992). Recently, researchers have begun developing comprehensive urban simulation models that can reflect the dynamics of population and urban environments (choices included) (Iacono et al. 2011).

As the demand for urban spatial evolution and simulation becomes more widespread in China, research on urban spatial simulation models has deepened. Wu and Webster (1998) proposed the first cellular automaton model in a Chinese urban context (Sim Land), incorporating different scenarios of AHP decision-making into the land use change simulation process (Wu 1998). Other scholars have constructed GIS-based cellular automaton models to simulate geographic spatiotemporal evolution processes. For instance, Sun (1999) proposed a CA model framework integrated with a Geographic Information System to vividly simulate and predict urban growth and diffusion in Ann Arbor, USA. Han et al. (2003) extended the CA model and integrated it with GIS, proposing a new method to simulate the spatiotemporal dynamics of urban growth and land value appreciation, analyzing the city of Deyang in Sichuan Province as a case study. Luo et al. (2003) organically integrated cellular automata with classic urban models, deriving a spatiotemporal model of urban population density based on homogeneous geographical backgrounds and the isolated city assumption, serving a fundamental role in controlling urbanization trajectories, to some extent compensating for the simplicity of CA modeling. Additionally, there have been numerous studies on improvements and expansions of the CA model (Ke and Bian 2010; Li and Liu 2007; Yang et al. 2015; Zhang et al. 2013).

These models can generally be classified into three types: traditional models, integrated models, and three-dimensional models.

Traditional CA models: These models appeared earliest and are the most widely used. The Centre for Advanced Spatial Analysis at University College London conducted early systematic research on the theoretical background, expansion mechanisms, and application methods of urban CA models, achieving fruitful research results and training influential scholars in the field of urban CA models, such as Michael Batty, Isaac Benenson, and Paul Torrens. These scholars extensively studied the dynamic evolution of urban expansion, land use, and residential patterns based on expanding standard CA models, forming a series of CA model elements and expansion mechanisms for buildings and urban spaces, significantly contributing to the development of urban CA models.

Integrated models: The complexity of spatial systems is difficult to describe with a single model, necessitating the expansion of the application capabilities of urban CA models through integration with other spatial models. The most common integration is between CA and Geographic Information Systems (GIS), leveraging CA's advantages in spatiotemporal dynamic modeling and GIS's powerful spatial data management capabilities for complementary benefits. Additionally, combining CA models with fractal algorithms can reflect the fractal characteristics of spatial systems, better predicting and controlling the evolutionary direction of spatial systems. Neural network models, genetic algorithms, and other methods can also be integrated with CA models to further enhance the application capabilities of urban CA models.

Three-dimensional models: Two-dimensional urban CA models mainly express planar spatial information and cannot fully describe the development of three-dimensional space. Given the reliance of contemporary architecture and urban dynamics on three-dimensional space, it is necessary to extend CA models to three dimensions to more accurately and realistically represent the characteristics of spatial systems. To date, research using three-dimensional CA models for spatial studies remains exploratory, but their importance is increasingly evident, making the three-dimensional development of urban CA models an inevitable trend.

Common Cellular Automata (CA) models include SLEUTH (Clarke and Gaydos 1998), MCE-CA (Wu 1998), Constrained CA (White et al. 1997), Logistic-CA (Wu 2002), IF–THEN Rule CA (Li and Yeh 2004), ANN-CA (Li and Yeh 2002), and CBR-CA (Li and Liu 2006), among others. The SLEUTH model summarizes urban spatial growth into four types: spontaneous growth, diffusive growth, edge growth, and road-influenced growth, providing researchers with a more intuitive understanding of urban evolution patterns. Models like MCE-CA, Constrained CA, and Logistic-CA express the probability of cells transitioning from a "non-urban" to an "urban" state as a function of urban development suitability and neighborhood effects. IF–THEN Rule CA and ANN-CA leverage machine learning methods to handle complex non-linear relationships, thereby improving the accuracy and reliability of model outcomes. The CBR-CA model circumvents the challenges of expressing the spatiotemporal heterogeneity of urban systems through rule-based approaches by using case-based reasoning, directly applying "old experiences" (i.e., a case library composed of sample data) to solve "new problems" (predicting whether a location transitions from a "non-urban" to an "urban" state), making it suitable for simulating urban growth on a regional scale with broad spatial extent and high heterogeneity (Chen and Li 2020). Although these models each have unique features and embody different approaches to modeling urban processes, their fundamental principle is to simulate global, complex urban development patterns through simple local transition rules, embodying the essence that "simple subsystem interactions can combine to form complex systems" (Li 2007).

Due to the inherent limitations of CA models and the complexity of urban systems, urban simulations based on CA face adaptability challenges. Land, as the bearer of human activities, inevitably reflects human consciousness in its development and evolution. In this sense, the growth, transformation, and decline of land are results of adaptation to the environment. In recent years, simulating the growth and transformation of land has been a hotspot in urban simulation (Batty 1994; Clarke et al. 1997; Wu and Webster 2000), but few studies have modeled urban decline, or counter-urbanization. Counter-urbanization results from the continuous degradation of land's adaptability to the environment, leading to urban areas gradually being outcompeted and phased out. Therefore, adaptability and control issues in simulation should be considered in realistic urban modeling. Traditional CA models generally do not account for macro factors, with system self-organization arising from local interactions among system elements. Control factors are singular, and state changes depend on the combination of the state of the cell itself and its neighbors. However, urban development and evolution depend not only on the action of local rules within the system but also on the results of multiple factors acting at various scales. Therefore, an effective and rational combination of micro-self-organization and macro influence factors should be adequately addressed in CA models.

3.2.3 Land Use and Transportation Interaction Models

1. **Concept and Basic Algorithms of LUTI Models**

The Land Use and Transportation Interaction (LUTI) models are computer models designed to simulate the endless development process of urban spaces using recursive algorithms. The term "land use" in "Land Use-Transportation Interaction" does not refer to the physical use of land in the conventional sense but rather to socio-economic activities that require space, typically places where people live and work. The focus on "land use" quantity is often on indoor space area. Some land use models are concerned with physical land use in the strict sense (Landis 2001), but these models are not referred to as "Land Use-Transportation Interaction models" (Simmonds and Feldman 2011) and thus are not discussed in this context. LUTI focuses on two main categories of urban socio-economic activities: households and firms (including various employment units). These economic activities require "indoor space" and transportation facilities, while transportation, in turn, can affect the location of socio-economic activities. The evolution of urban spaces is a continuous interaction process between the land use system and the transportation system (Lowry 1964).

Existing LUTI models generally share a similar architecture, consisting of transportation and land use models. The transportation model evaluates urban transport accessibility based on the spatial distribution of urban activities and the transportation network. In contrast, the land use model predicts urban spatial development trends based on transport accessibility and other influencing factors. As socio-economic activities' spatial distribution changes under the influence of transportation, it inevitably alters factors such as rents and accessibility in urban locations, leading to changes in the spatial distribution of socio-economic activities and subsequent adjustments in rents and other factors. This mutual interaction continues until equilibrium (or a certain state) is reached. Figure 3.14 provides a general structural framework containing recursive algorithms. The specific implementation details of the transportation and land use models (location models) vary across different cases, leading to functional and technical differences in LUTI models.

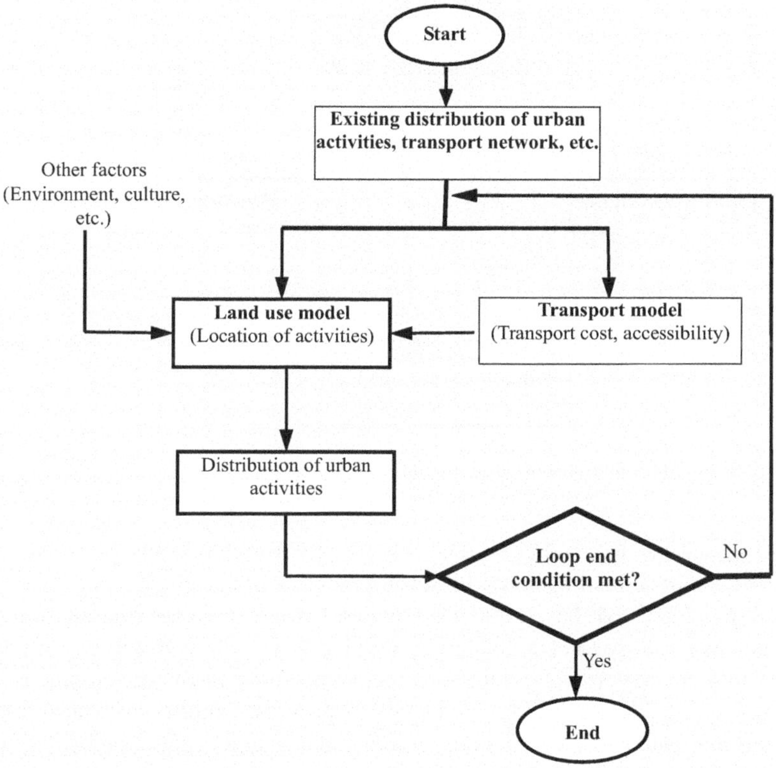

Fig. 3.14 General process of LUTI model algorithm

2. The Origin and Development of the LUTI Model

LUTI models originated in the 1950s when American scholars first systematically studied the mutual relationship between urban transportation and spatial development. Hansen's research suggested that locations with better transportation had better development opportunities and higher densities; urban transport and socio-economic activity locations influence each other, requiring coordinated and harmonious planning decisions (Hansen 1959). This idea became popular in the American planning community, laying the foundation for the LUTI theoretical framework. The interaction between urban land use and transportation can be summarized as follows: The spatial distribution of urban land uses (such as residential, industrial, and commercial areas) determines the spatial separation of human activities (living, working, shopping, etc.), which interact through transportation, while transport accessibility (accessibility) determines the choice of activity locations, in turn, causing changes in the land use system, and so on in a cycle. The feedback process between the land use system and the transport system is illustrated in Fig. 3.15.

Subsequently, in the early 1960s, quantitative simulation received unprecedented attention in the field of urban planning. To achieve urban quantitative simulation and

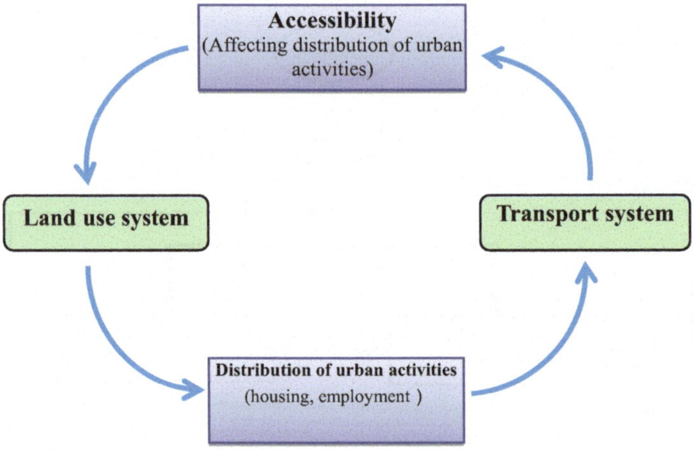

Fig. 3.15 Urban land use and transportation mutual feedback system

assist decision-making, researchers extensively adopted simulation analysis methods from other disciplines, leading to rapid development in urban models and simulation technologies. In 1964, based on LUTI theory, Lowry made the first attempt to use computer technology to establish an urban model, known as the Lowry model (Horowitz 2004; Lowry 1964), which became a milestone in the development of the LUTI model. The Lowry model views urban space as composed of transportation networks and land use. Here, land use refers to the spatial occupation by various economic activities, including residential, production, and service sectors. The model posits that the urban spatial development process is an interaction of various economic activities through transportation, with the foundational production sector being the basis for the city's existence. The goods or services produced exceed the city's boundaries, sold to external markets, bringing income to the city; whereas the goods and services produced by the service sector are consumed within the city and are essential for its operation. Given the location of the basic sectors, the Lowry model outputs the spatial distribution of residential areas and service sectors (i.e., urban land use). The Lowry model includes residential location models and service sector location models, with models nested within each other. The model operates as follows:

(1) First, determine the spatial location of the production sectors. The spatial location of the production sectors, determined by transportation, decides the residential locations of their workforce, which the model considers as determining the city's total population.
(2) The location of residents determines the commercial (service) value of the surrounding plots, thus determining the spatial distribution of various service sectors.
(3) The population of the service sector also requires residential land, with their residential locations determined by the spatial locations of the service sectors. This changes the total population, around which the value of plots changes,

generating more service demands (new service sector personnel also require services).

(4) Repeat steps 2 and 3 until the change in total population is minimal. At this equilibrium state, the city's spatial development tends toward this state.

Given the spatial distribution of basic sectors and the transportation network, the Lowry model outputs the spatial distribution of residential areas and service sectors (i.e., the condition of urban land use). The interrelationships of spatial locations among production sectors, residential areas, and service sectors lay the conceptual framework for the LUTI model, with subsequent model developments based on this approach. The Lowry model, being a computer simulation model, requires iterative processing by computer programs to simulate the urban spatial development process. The model runs continuously until the results stabilize, i.e., until changes between two iterations are minimal. However, the actual urban development process, including the scale of economic activities and transportation infrastructure, is dynamic and evolving, which has become an important area of expansion for later developments of LUTI models. Additionally, the Lowry model is static, depicting the city's development trend at a certain moment, uses displacement distances to represent traffic conditions between locations, and evaluates the attractiveness of plots based on market principles. These aspects have been the main focus of subsequent scholars' extensions. Scholars have extensively applied and expanded the Lowry model, with many models commonly used in urban planning in Europe and America in recent years based on its theoretical framework.

Currently, there is no unified understanding or standard for classifying LUTI models. For ease of understanding and interpretation, scholars have categorized LUTI models from different perspectives and analyzed their developmental trajectories. Miller et al. (1998) categorized transportation models into four types based on their developmental process:

(1) single-mode transportation models (T1), simulating car travel only;
(2) multimodal public transportation models (T2);
(3) logistic-based transportation models (T3), capable of simulating traffic flows at different times;
(4) micro traffic flow models based on economic activities (T4).

Land use models were divided into five types based on their development:

(1) stages without land use models (L1);
(2) determining the location of economic activities based on occupational nature (L2);
(3) non-market land allocation models (L3);
(4) market price-based land allocation models (L4);
(5) market-based economic models (L5).

Miller combined these land use and transportation model developments into a matrix. Later, Wegener added another land use model (L6) (Wegener 2004), combining it with Miller's classification to form a new matrix as shown in Table 3.10.

The matrix describes the developmental process of LUTI models, where each cell represents a combination of a type of land use model and a type of transportation. The direction of the arrows indicates the gradual enhancement of LUTI model functions.

In addition, other well-known classification methods include:

(1) Based on the relationship of variables within the model, they can be divided into static models and dynamic models (DSC/ME&P 1999; Niu et al. 2014). Static models characterize the urban spatial evolution trend under the assumption of constant input variables at a single time point. Most early LUTI models were static, with the most classical being the Lowry model (Chen et al. 2010). Subsequently, many scholars proposed a series of static models, such as DSCMOD, MUSSA (DSC/ME&P 1999), etc. Since any change in conditions during the city's spatial development process will be reflected after a period of time, and it's challenging to keep all element conditions constant over time, the results of static models represent an idealized equilibrium, indicative only of current trends. Dynamic models simulate a series of time periods, where the spatial development state of a certain period is affected by the variables of the previous period. In other words, adding a time axis to static models, where various input variables change over time. Dynamic models can further be divided into three categories (DSC/ME&P 1999): I entropy models, which are based on statistical analysis methods, initiated by Alan Wilson, such as the LILT model (Mackett 1983); II spatial economic models, which integrate economic models and spatialize them, such as MEPLAN, TRANUS, METROSIM; III activity-based models, which describe the interaction process between different economic activities, such as DELTA, URBANSIM, IRPUD models.

(2) Scholars like Pierluigi have classified LUTI models into three generations and five categories based on the development process and implementation methods of the model: The first generation models appeared in the 1960s and 1970s,

Table 3.10 The evolution of the LUTI model (Miller et al. 1998)

Transport Model / Land use model	T1	T2	T3	T4
L1				
L2				
L3				
L4				
L5				
L6				

Note L6 was proposed by Wegener (2004)

including spatial interaction models, mathematical planning models, and input–output models; the second generation models were the random utility models that emerged from the 1970s to the 1990s; the third generation models are the micro-simulation models developed after the 1990s (Pierluigi et al. 2013).

(3) Chang classified LUTI models into four major categories based on the model's mathematical structure: spatial interaction models, mathematical programming models, random utility models, and bid rent models (Chang 2006).

(4) Waddell classified LUTI models based on the model's theoretical foundation (Waddell and Ulfarsson 2004).

These classifications provide a macroscopic description of the models' characteristics and development trajectories, aiding in understanding and using the models. However, due to the diverse and complex structure of LUTI models, it's challenging to categorize all models within a single classification system (Chang 2006). To further explore the implementation of LUTI models, this document will discuss the implementation of each component (sub-model) of the model.

3. **Common Technical Methods and Modeling Principles of LUTI Models**

(1) **Common Technical Methods of LUTI Models**

LUTI model technical methods refer to mathematical methods used to assess the intensity of spatial interactions or predict locations. The key mathematical principles summarized below describe the simulation mechanisms of most models.

(i) **Assessing Spatial Interaction Intensity**

Spatial interaction serves as the driving force of LUTI (Land Use and Transport Interaction) models. The focus of LUTI on spatial interactions primarily involves flows between urban zones, such as logistics and human flow, with a particular emphasis on human flow (travel). The model needs to employ various indicators to assess the traffic flow between urban zones, for example, predicting the traffic flow of work travel based on the spatial distribution of population and employment activities and the cost of transportation. The commonly used model for this purpose is the gravity model, which is based on the assumption of spatial interaction's similarity to Newton's law of universal gravitation. According to Newton's law, the gravitational force between two bodies is directly proportional to the product of their masses and inversely proportional to the square of the distance between them. By applying this formula to geography, the flow between regions (e.g., the number of commuters) is considered as an interaction force between regions, with mass variables replaced by structural variables, such as population, GDP, etc., thereby representing the regions' capacity to generate flow. The use of the gravity model entails an important assumption: the size of flow between regions is directly proportional to the scale variables of two points and inversely proportional to the square of the distance. However, there is currently no theory to prove that traffic flow between regions is inversely proportional to the square of the distance. Therefore, it is more meaningful to change the denominator in the calculation of interaction intensity between regions to the nth power of distance, where the value of n should vary according to demand. This leads to the most basic model for evaluating spatial interaction. This model can be interpreted as the traffic flow between regions being caused by the push and pull of the two regions, influenced by distance.

The gravity model describes the relative distribution of interaction intensity between regions. In practice, it often necessitates imposing constraints on both ends of the interaction, that is, compromising with known information, such as the number of residents at the starting point and the number of jobs at the destination. This requires applying constraints to the gravity model to form a constrained gravity model (Thomas and Huggett 1980).

Furthermore, entropy theory has also been introduced into the study of spatial interaction, used to solve for the probability of the system reaching a certain state. When data is insufficient, it's necessary to use available information to solve for the system's potential states, and entropy maximization is an important method for this. Entropy theory includes two crucial concepts: macro-states and micro-states. The entropy maximization solving process is based on solving the most probable macro-state from micro-states, for example, given the number of trips in a city, solving for the OD (Origin–Destination) flow between pairs of regions can adopt the principle of minimizing total transportation costs (Barra 1989). Similar to the gravity model, the entropy model can also be transformed with different constraints applied.

(ii) **Location Simulation**

The models concerning spatial interaction are primarily utilized for simulating OD (Origin–Destination) flows, where constrained models can be applied to simulate location choices, such as predicting travel space distribution based on the number of residents. Additionally, location choice simulation typically involves discrete choice models and logit models.

Discrete choice models are often used for explaining phenomena to provide decision support. Common discrete choice models include the random utility model and its derivatives. The random utility model is based on a series of assumptions: (1) decision-makers are faced with a set of discrete, independent choices—to choose or not choose a particular option; (2) the principle of decision-making is utility maximization; (3) the likelihood of choosing a specific option is represented as a probability; (4) the utility of each decision can be divided into two parts, measurable utility and random utility, where the former can be evaluated based on selected indicators, and the latter refers to unknown factors that disturb the evaluation result (Golledge and Stimson 1997). Different weights can be added to vary the model.

The logit model is derived from the assumption of random utility to extend the random utility model. The logit model includes multinomial logit models and nested logit models. McFadden's logit model posits that the probability of a decision-maker choosing option k is determined by the characteristics of the options and the characteristics of the decision-maker themselves (Government of Ireland 1995). McFadden (1978) applied the Weibull distribution function to the random utility model, deriving the logit model that is widely used today. This model assumes that decision-makers will evaluate each option before making a decision, but in reality, due to incomplete information, it is difficult for decision-makers to weigh every choice in a city individually, and they only assess a subset of options. Moreover, structurally, the logit model does not consider the interrelationships between options. In reality, various spatial options are often interconnected. Based on these considerations, nested logit models (Nested logit models) have emerged, which are not detailed here.

(iii) **Rent Theory**

Another important simulation technique in LUTI models is rent theory. Rent theory, systematically proposed by Alonso (1964), is similar to von Thünen's agricultural location theory (Thunen and Heinrich 1826). Rent theory explains the spatial distribution of urban activities based on a series of assumptions. The core idea of rent theory is that the rent of each location is determined by its quality, profit, and utility, which are usually dependent on the distance from the location to the city center. Although this theory has been widely applied in urban simulation, its theoretical foundation is limited. Its utility function, which often emphasizes economic factors (transportation costs), is typically difficult to realize, as many non-economic factors also play a role, such as air quality, safety, comfort, etc. (Torrens 2000). As a variant of rent theory, the hedonic price model further decomposes housing prices into several components (such as land quality, structure, number of bedrooms, environment, etc.), but obtaining detailed price data, especially over many years, is challenging.

(2) **Principles for Constructing LUTI Models**

There has been significant accumulation of research on LUTI models, leading to the establishment of various model frameworks and the development of corresponding general-purpose software tools. Thus, suitable modeling tools can be selected based on the characteristics of different cities to construct specific application models. To develop proprietary simulation software, new models must be constructed from the ground up, and simulation software developed based on these models. Beyond theoretical depth, operability, and practicality, model construction should consider the following requirements:

(1) The model should reflect the interaction between land use and transportation supply, such as the impact of land use changes on transportation and the impact of changes in transportation facilities on land use (distribution of economic activities).
(2) The model's inputs should reflect the proposed plans of planners, including suggestions for spatial development in land use, transportation, etc., thus facilitating experimentation with planning schemes and policies.
(3) The model's outputs should address relevant policy questions, enabling the assessment of alternative scenarios, such as evaluations of the sustainability of land use and transportation policies, socio-economic impacts, and environmental effects.
(4) The model should take into account the main growth trends in urban space and the corresponding measures for evaluating land use and transportation policies, capable of simulating changes in urban areas.

(5) The model should be well-supported by existing data to ensure good calibration and validation. Before applying the model, it should be tested with existing data to determine its suitability for decision support.
(6) The model should provide answers within a suitable time frame to facilitate policy analysis.

When constructing LUTI models in different countries or regions, besides considering the common requirements of the model, the specific political and cultural background of each must also be reflected in the model, either as operational constraints or input parameters, incorporating as many influencing factors as possible to produce more reliable predictions.

(3) **Discussion on Constructing LUTI Models for Chinese Cities**

Under the backdrop of globalization, China is experiencing rapid urbanization, with urban spatial development driven by both internal and external, even global, influences, adding complexity to spatial decision-making. Effective policy evaluation tools are urgently needed to ensure the scientific nature of urban spatial policies and promote the healthy development of cities. Thus, constructing LUTI models suitable for the spatial development of Chinese cities has significant practical importance. However, quantitative simulation and systemic analysis in urban research are still limited in China, with few examples of city-level comprehensive models in urban planning practices. Conceptually, urban development in Chinese cities is supported by three phases of planning: separate planning processes (land use planning and transportation planning), integrated land use and transportation planning, and sustainable development planning (land planning, transportation, and environment). Current practices in urban modeling are still in the first phase, far from full-scale modeling.

However, China has its unique historical and cultural background and political context, differing from the social development stages of foreign developed countries. Therefore, while foreign experiences can be referenced, they must be integrated with China's realities. The construction of LUTI models for Chinese cities should consider: (1) the uniqueness of China's political and cultural background, such as administrative divisions and cultural diversity, which are unavoidable factors affecting the effectiveness of policy implementation, differentiating it from the implementation of urban spatial policies abroad; (2) the stage of socio-economic development, though China's market economy system continues to improve, traces of the planned economy still exist, and the government's macro control over urban spatial development, such as population distribution and infrastructure construction, cannot be ignored; (3) China's household registration system, which significantly impacts population mobility and residence, is a crucial simulation content of LUTI models. Additionally, model calibration is a necessary step in the modeling process to ensure the effectiveness of LUTI model simulations. Utilizing historical data over many years for calibration, adjusting parameters based on errors, and repeatedly calibrating to obtain optimal parameter values are essential.

4. **Application of LUTI Models in Urban Spatial Evolution Simulation**

(1) **Problems that LUTI Models Can Help Assess**

The core function of LUTI models lies in their ability to predict future urban spatial development trends based on policy parameters, with sustainability assessments being conducted based on these predictions. Using LUTI model simulations, the following key issues related to sustainable development can be addressed:

(1) Quantitative assessment of land resource occupation. By considering the spatial distribution of the population and various economic sectors, the occupation of land resources can be calculated further. Urban expansion may encroach on a significant amount of arable land, potentially leading to serious ecological and environmental issues. Therefore, scientifically quantifying the impact of different urban spatial policies on land resource usage is a critical consideration for decision-makers.
(2) Energy consumption. The model can simulate the consumption of energy sources such as gas and gasoline in the city, based on the spatial distribution of the population and economic sectors and traffic flow conditions. As urbanization and the city's economy continue to grow, the energy consumed by the city will further increase. The consumption of energy is an indispensable issue in sustainable development.
(3) Emissions of waste gases and pollutants. Economic development leads to increased energy consumption and land occupation, followed by an increase in emissions of waste gases and pollutants. The emission of these pollutants can harm the ecological environment and degrade the quality of life. Sustainable development requires reducing the emission of these pollutants and enhancing the effective treatment of emissions.
(4) Economic condition assessment. The implementation of urban spatial policies will inevitably affect economic development. Based on the spatial distribution of the population and economic sectors, simulations can assess economic conditions such as traffic situations and market vitality.
(5) Social development condition assessment. The condition of social development is a deeper policy effect issue. During the formulation of urban spatial policies, how to promote healthy social development is an unavoidable question. Simulations can assess employment conditions, education levels, connections between regions (residents), and social equity based on the distribution of the population and various economic sectors, as well as traffic conditions.

The conceptual model shown in Fig. 3.16 for policy testing using the LUTI model, where the "Land Use-Transport Interaction System" is the core part of the framework, i.e., the specific implementation part of the LUTI model. The model should be able to incorporate spatial policy parameters. According to Fig. 3.16, LUTI inputs include the current state of urban spatial development and spatial policies, while outputs include future land use conditions. Based on the model's outputs, the sustainability of urban development can be evaluated, thereby testing spatial development policies. Spatial policies can be categorized into three main types: investment and construction, charges, and regulatory measures.

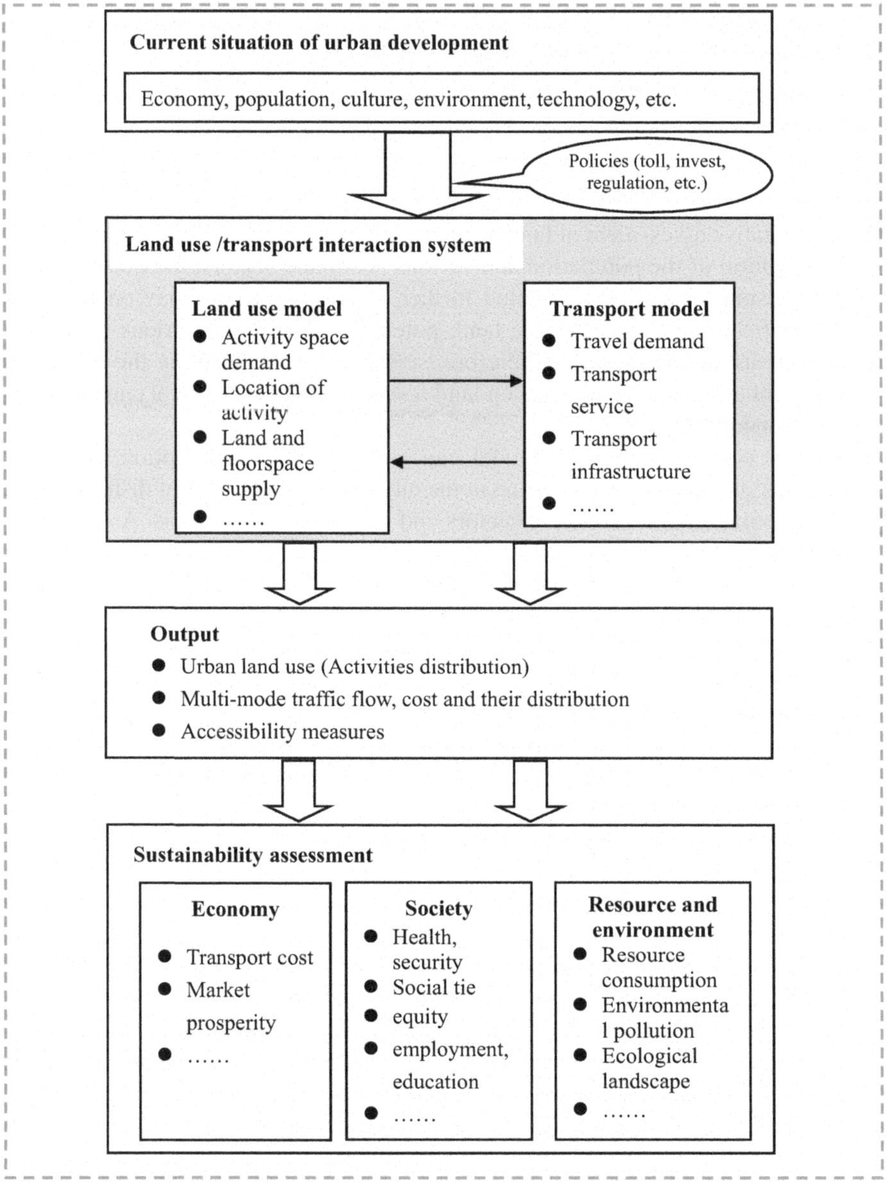

Fig. 3.16 Conceptual model for policy testing using the LUTI model

According to Fig. 3.16, the work of simulating urban spatial development and assessing sustainability is divided into three steps: first, conducting scenario analysis on the current state of urban spatial development and development strategies to determine the implementation policies of different spatial strategies; second, establishing the LUTI model to describe the spatial behavior of land consumers such as the population and economic sectors under the interaction of transportation and land use (including travel and site selection), predicting future land use and traffic conditions (model output items); third, evaluating the output results to determine the sustainability of spatial policies.

(2) **Application of LUTI Models in Assisting Urban Spatial Decision-making Abroad**

To assist spatial decision-making and achieve sustainable development of cities and regions, scholars abroad have conducted extensive research on the extension and application of LUTI models. With the aid of computer software technology, simulation application systems have been established and applied successfully in practice, providing solid support for the formulation of urban spatial policies. For example, to curb urban sprawl, the UK implemented a series of policies, including establishing green belts around cities, successfully limiting urban sprawl and protecting agricultural land. However, on the other hand, limited land supply led to rising land prices, causing urban residents' living spaces to shrink compared to other European countries; rising house prices also affected some businesses' affordability, reducing their competitiveness; even more, some city residents moved beyond the protected areas to live in villages or small towns, increasing their commuting distances and leading to traffic congestion, exhaust emissions, and traffic accidents. To solve these problems and achieve sustainable urban development, decision-makers face three strategic choices: compact strategy, market-oriented free expansion, and planned urban expansion. To choose a more sustainable development strategy, it is necessary to simulate and evaluate the future conditions and sustainability of various development options. For this purpose, Echenique and others used LUTI models for urban regions in Tyne, Cambridge, and London as case studies, set different scenarios, and evaluated the sustainability of each development strategy, concluding that urban form is not a decisive factor (Echenique et al. 2012; Gordon et al. 2011), and compact city forms are not necessarily more sustainable, providing recommendations for implementing sustainable spatial strategies. In the 1980s, the UK Department of Transport developed the LASER (London and South East Region) model based on MEPLAN software (Echenique 1994; Willianms 1994), which was used to simulate the spatial distribution of population and employment in London and the southeast region for 1981 and 1991, as well as traffic flows for different transportation modes, purposes, and routes, providing a basis for sustainability assessments of urban development. The LASER model is a typical example of the application of LUTI models in sustainable development simulation analysis. The LASER model, which has been tested and revised over a long time and developed into its third version by 2009 (Echenique et al. 2012; Feldman et al. 2009), is still used by the UK government to evaluate the effects of different spatial policies, supporting the formulation and implementation of sustainable development policies in London and its surrounding areas. Feldman and others established the Auckland LUTI model, and Dobson and others developed the Manchester area LUTI model (Dobson et al. 2009). Other countries also have numerous successful application cases: Australia's TOPZA, TRACKS, TRANSTEP, TOPMET; the United States' IT-LUP; the Netherlands' AMERSFOORT; Japan's CALUTAS, OSA-KA; Germany's IRPUD; Sweden's SASLOC; Venezuela's TRA-NUS, all of which have played an active role in urban or regional planning (DSC/ME&P 1999; Oryani et al. 1997).

(3) **Application of LUTI Models in Urban Spatial Decision-making in China**

As early as around the year 2000, domestic scholars had already paid attention to the Lowry model, mainly limited to its introduction and discussion. Scholars like Yang Wuyang, Wang Jixian, and Zhou Suhong introduced the concept and thought of the Lowry model through a review of foreign land use models (Wang 2001; Yang and Jinshe 1997; Zhou and Yan 2005). Following this, some scholars began to try improving and applying the model. Chen Peihong and others discussed improvements to the form of the Lowry model (Chen and Wang 2007). Liang Jinshe and others analyzed Beijing's urban expansion based on the Lowry model, concluding that the development of Beijing indeed exhibited factors considered by the Lowry model (Liang and Chu 2005). Zhou Binxue and others utilized the Lowry model's theoretical framework about the division of economic sectors to study urban spatial structure, focusing on the interactions between various urban elements (transport, location, land use) and using known data to calculate the spatial distribution of unknown elements. Their research initially applied concepts from Lowry's theory but did not involve predicting future urban development trends, hence the recursive thought of the Lowry model was not applied (Zhou et al. 2011, 2013). It is evident that although there is interest in LUTI models in China, research remains at the stage of understanding, organizing, and theoretically exploring the early Lowry model (Zhou et al. 2013), with few application studies limited to referencing its conceptual framework. The core function of LUTI models for simulating urban and regional spatial evolution and predicting development trends has not been applied.

The reasons can be summarized as follows:

(1) Domestic urban research, especially quantitative research, started late, and currently, few scholars are involved in LUTI model research;
(2) Domestic research has traditionally focused on macro issues and qualitative analysis, and although quantitative simulation research has received increasing attention, comprehensive integrated simulation studies are still weak;
(3) The LUTI model is essentially a theoretical framework for a computer model (Lowry 1964), using the computer's endless loop processing capability to simulate the city's endless development process. Model development requires not only expertise in urban development and economics but also computer programming skills, hindering the widespread application of LUTI models;
(4) The operation of LUTI models requires a large amount of detailed urban land use, transportation data, and extensive mathematical processing, limiting their

widespread use. The development and application of the core function of LUTI models for simulating the urban spatial evolution process in China are still weak, presenting opportunities and challenges for research in this area.

5. **Limitations and Development Prospects of LUTI Models**

(1) **Limitations of LUTI Models**

(i) **Theoretical Level**

Simulating the urban spatial evolution process involves many fields, making it a systemic and complex task. LUTI models, developed for this purpose, are still considered one of the most effective tools for long-term urban planning. However, compared to the recent speed of urban development, the development of LUTI theory has been somewhat lagging. This is primarily because current model development focuses more on addressing immediate real-world problems rather than the theoretical relationships between urban land use and transportation described by the models.

Firstly, the assumptions or theoretical bases upon which LUTI relies were established in an earlier era (the 1960s), when cities were very different from today. They mainly focused on spatial interactions, considering location as the primary driver of urban land use and transportation development. However, besides location, there are many other significant influencing factors. Secondly, current LUTI models mainly use an aggregated approach to analyze urban spatial interactions, lacking attention to interactions between individual entities within cities. There is a lack of multi-level simulation from micro-individuals to macro-phenomena. Furthermore, current LUTI models are usually static or quasi-dynamic, meaning the model converges to an equilibrium state. However, urban development is often seen as a dynamic process with continuously changing factors, never reaching an equilibrium state. Therefore, how to further align these aspects remains an area for exploration. Short-term predictions using these models may have large errors, while long-term predictions can be more meaningful as urban spaces, although not reaching a specific equilibrium state, tend to approach a possible equilibrium.

Due to the aforementioned theoretical shortcomings, current applications of the model are still limited to addressing some traditional questions, such as the impact of land use planning or real estate development on the land use system and transportation, and how changes in transportation affect the spatial distribution of urban socio-economic activities.

(ii) **Application Level**

Main shortcomings in LUTI case applications include: large spatial scales that are not conducive to simulating socio-economic behaviors and the effects of spatial policies at the community scale; traditional four-stage approach used in transportation models being inadequate for experimenting with the impact of spatial policies on human behavior; weak integration of environmental quality assessment models; reliance on utility or profit maximization for simulating urban development, focusing on economic laws while neglecting social phenomena like urban spatial fractality or polarization; and the primary research and application of LUTI models being in developed countries, with scarce application in urbanizing developing countries like China.

There is no consensus on which model outputs are better or more reliable. Essentially, LUTI models are used to describe the interrelationship between urban socio-economic activity locations and transportation, so the reliability of models should be evaluated based on this fundamental structure.

(2) **Development Prospects of LUTI Models**

At its core, LUTI theory explains the interrelationship between urban land use and transportation. Observing the model's development process from its inception to the present, the theoretical framework has not undergone substantial improvements. There is an anticipation for LUTI models to achieve further breakthroughs in theory, methodology, and application.

The model is expected to expand more towards a micro-level. It is believed that current model implementations are predominantly top-down, whereas urban activities are actually composed of various micro-individuals. These individuals exhibit spontaneous and disordered movements that manifest as regular macro phenomena. Therefore, a significant advancement would be to start from the bottom-up, simulating individual behaviors to capture urban development laws. Currently, simulations based on Agent technology are more conducive to micro-level modeling. Agent technology is often used in conjunction with Cellular Automata (CA), gradually shifting from physical land use simulation to focusing on the relationship between urban land use and transportation, thereby converging agent-based CA models with LUTI models. The rapid advancement of computer technology provides favorable conditions for high-resolution simulations, setting the stage for theoretical breakthroughs.

LUTI focuses on internal spatial changes within cities and lacks connection to macro issues. Firstly, it is challenging to evaluate the impact of spatial policies on the city's macroeconomic system. As urban development is inevitably influenced by regional macroeconomics, deepening the integration with regional macroeconomics awaits further exploration. This could also facilitate the broader application of the model in developing countries. Secondly, current models appear weak in studying spatial equity, social group clustering, exclusion, and the resulting social issues. Urban spatial structure determines spatial equity, such as fairness in education, healthcare, and environmental aspects. Combining the spatial distribution of socio-economic activities with equity evaluation and further theoretical expansion has significant practical implications. Additionally, in the context where sustainable development becomes another focal point, integrating sustainable development simulation analysis functions presents a new task for LUTI model development, necessitating enhanced environmental and ecological assessments at the application level.

On the operational level, existing LUTI model applications are relatively complex, demanding high levels of professional knowledge and computer literacy from users.

For broader application, model development needs to consider universality, operability, data requirements, model validation, and stability. Overly complex models are difficult to operate and bring about issues in validation, stability, and data preparation. Therefore, constructing LUTI models requires balancing theory and methodology with model functionality and operability. Additionally, with the rapid development of visualization technologies like GIS, incorporating LUTI models with such technologies can facilitate broader user participation in urban simulation.

Regarding the expansion and application of LUTI models in China, the current urbanization context demands scientifically sound land use and integrated transportation policies. Introducing LUTI theory into China for urban land use and integrated transportation simulation research to assist decision-making is a challenge scholars must address. Inherently, LUTI models articulate the essence of the interplay between urban land use and transportation, indicating no specificity in the application of LUTI theory to Chinese cities. The difference lies in the model's functionality selection, parameter settings, and calibration for different cities. However, obtaining the core codes and specific implementations of foreign models is generally challenging. Chinese scholars can leverage the latecomer advantage, facing the challenges head-on while drawing on foreign experiences to advance LUTI theory development. Conducting LUTI model application research will further enrich and develop the theory and methodology of urban research domestically.

3.3 Chapter Summary

The academic community has long focused on urban space, accumulating a wide array of research methods commonly referred to as urban models or simulation methods. However, there is a significant functional disparity among these models. Based on their functions, they can be categorized into two main types: urban analytical models and simulation models. Urban analytical models are typically used to assist researchers in understanding the laws of urban spatial development, characteristics, influencing factors, and mechanisms, and to comprehend or describe the interactions between various urban elements. These models are often specific mathematical models, such as regression analysis models (including linear and nonlinear regression), which help us understand the factors influencing urban development and their mechanisms, while complex network analysis methods help us understand the characteristics of urban transportation. Analytical models are based on historical data analysis, whereas simulation models fit the process of urban spatial development, abstractly expressing the development process of entities. Their primary function is to fit and predict urban development trends. Implementing simulation models first requires a comprehensive understanding and abstraction of the urban development process, i.e., establishing a conceptual model, followed by mathematical realization. The simulation process usually involves multiple mathematical models and can utilize different mathematical models for implementation. For example, the Lowry model (Lowry 1964) abstractly describes the urban spatial evolution process as a cyclic operation, well-suited for fitting with computer recursive algorithms, laying the foundation for simulating the urban spatial evolution process using computer technology. However, the specific implementation of the Lowry model involves numerous mathematical formulas, and the realization of each module using mathematical formulas is not unique, with continuous improvements made by later researchers.

Analytical models enable us to understand the laws of urban spatial development, representing empirical and inductive analysis of these laws, which is a prerequisite for simulation. Simulation models abstract and deduce the laws of urban system development. As quantitative research develops domestically, empirical analysis continues to improve, making the pursuit of simulation research for further urban spatial decision support unavoidable. Combining empirical analysis with simulation predictions will also lead to the refinement of urban studies.

References

Ahlfeldt GM, Feddersen A (2017) From periphery to core: measuring agglomeration effects using high-speed rail. J Econ Geogr 18(2):355–390

Ahmed M, Mario C, Ismail S et al (2017) Coupling agent-based, cellular automata and logistic regression into a hybrid urban expansion model (HUEM). Land Use Policy 69:529–540

Albalate D, Bel G (2012) High-speed rail: lessons for policy makers from experiences abroad. Public Adm Rev 72(3):336–349

Albert R, Jeong H, Barabási A-L (2000) Error and attack tolerance of complex networks. Nature 406:378–382

Aljoufie M (2014) Toward integrated land use and transport planning in fast-growing cities: the case of Jeddah, Saudi Arabia. Habitat Int 41:205–215

Alonso W (1964) Location and land use. Harvard University Press, Cambridge, MA

Amaral S, Monteiro AM, Câmara G et al (2006) DMSP/OLS night-time light imagery for urban population estimates in the Brazilian Amazon. Int J Remote Sens 27(5):855–870

Amormso S, Patt Y (1972) Decision procedures for surjectivity and injectivity of parallel maps for tessellation structures. Comput Syst Sci 6:448–464

Bagler G (2008) Analysis of the airport network of India as a complex weighted network. Physica: Statal Mech Appl 387:2972–2980

Banister D, Berechman Y (2001) Transport investment and the promotion of economic growth. J Transp Geogr 9(3):209–218

Barra TDL (1989) Integrated land use and transport modelling: decision chains and hierarchies. Cambridge University Press, Cambridge

Barthélemy M (2011) Spatial networks. Phys Rep 499:1–101

Batty XY (1994) From cells to cities. Environ Plan B: Plan Des 21(1):31–48

Batty M (2008) Fifty years of urban modelling: macro statics to micro dynamics. In: Albeverio S, Andrey D, Giordano P et al (eds) Dynamics of complex urban systems: an interdisciplinary approach. Physica A, Heidelberg, pp 1–20

Batty M, Vargas C, Smith D et al (2013) SIMULACRA: fast land-use-transportation models for the rapid assessment of urban futures. Environ Plann B 40:987–1002

Baum-Snow N, Brandt L, Henderson JV et al (2017) Roads, railroads and decentralization of Chinese cities. Rev Econ Stats 99(3):435–448

Beck T, Levine R, Levkov A (2010) Big bad banks? The winners and losers from bank deregulation in the United States. J Financ 65(5):1637–1667

Brandi A, Gori S, Nigro M et al (2014) Development of an integrated transport-land use model for the activities relocation in urban areas. Trans Res Procedia 3:374–383
Campos J, Rus G (2009) Some stylized facts about high-speed rail: a review of HSR experiences around the world. Trans Policy 16(1):19–28
Cao X (2010) Simulation and early warning study of urban land use change based on CA in Shenzhen City. Nanjing University. (in Chinese)
Cao Z, Wu Z, Kuang Y et al (2015) Calibration and application of DMSP/OLS night light images in China. J Geo-Inf Sci 17(09):1092–1102 (in Chinese)
Cecchini A (1996) Urban modelling by means of cellular automata: generalized urban automata with the help online (AUGH) model. Environ Plann B Plann Des 23(6):721–732
Chang JS (2006) Models of the relationship between transport and land-use: a review. Transp Rev 26(3):325–350
Chang Z, Murakami J (2019) Transferring land use rights with transportation infrastructure extensions: evidence on spatiotemporal price formation in Shanghai. J Trans Land Use 12(1):1–19
Chao Y, Li Q, Chen S (2007) Location modeling based on multi-agent system. Geomats Inf Sci Wuhan Univ 7:646–649 (in Chinese)
Chen S (1999a) Theoretical study on urban growth and land value increment. Geo-Inf Sci 01:12–18 (in Chinese)
Chen S (1999b) Urbanization and urban geographical systems. Science Press, Beijing (in Chinese)
Chen Y, Li X (2020) Applications and new trends of machine learning in urban simulation research. Geomat Inf Sci Wuhan Univer 45(12):1884–1889 (in Chinese)
Chen P, Wang J (2007) Transport and land use models: Theoretical foundation and improved forms of the Lowry model. Prod Res 14:77–80 (in Chinese)
Chen G, Lu G, Wang H (2000) The development and problems of urban model. Econ Geogr 20(5):59–62 (in Chinese)
Chen J, Gong P, He C et al (2002) Assessment of the urban development plan of Beijing by using a CA-based urban growth model. Photogramm Eng Remote Sens 68(10):1063–1071
Chen Y, Li S, Li X et al (2010) Simulating compact urban form using cellular automata (CA) and multi-criteria evaluation: A case study in Dongguan. Acta Sci Nat Univ Sunyatseni 49(6):110–114 (in Chinese)
Chen Y, Wang J, Jin F (2020) Robustness of China's air transport network from 1975 to 2017. Physica Stat Mech Appl 539:122876
Clarke K, Gaydos L (1998) Loose-coupling a cellular automaton model and GIS: long-term urban growth prediction for San Francisco and Washington/Baltimore. Int J Geogr Inf Sci 12(7):699–714
Clarke KC, Gaydos LJ, Hoppen S (1997) A self-modified cellular automaton model of historical urbanization in the San Francisco Bay area. Environ Plann B Plann Des 24(2):247–261
Converse PD (1949) New laws of retail gravitation. J Mark 14:379–384
Coppola P, Ibeas A, dell Olio L et al (2013) LUTI model for the metropolitan area of Santander. Urban Plann Dev 139(3):153–165
Couclelis H (1985) Cellular worlds: a framework for modeling micro-macro dynamics. Environ Plan A 17(5):585–596
Couclelis H (1989) Macrostructure and micro behaviour in a metropolitan area. Environ Plan B 16:151–154
Crespo A, Garcia-Molina H (2000) Modeling archival repositories for digital libraries. In: Proceedings of the 4th European conference on research and advanced technology for digital libraries. Springer-Verlag
Cui X, Fang C, Zhang Q (2018) Spatial relationship between high-speed transport superiority degree and land-use efficiency in Shandong Peninsula urban agglomeration. Acta Geogr Sin 73(6):1149–1161 (in Chinese)
Dai E, Ma L, Yang W et al (2019) Agent based model of land system: Theory, application and modelling framework. Acta Geogr Sin 74(11):2260–2272 (in Chinese)

Deadman PJ, Brown RD, Gimblett HR (1993) Modelling rural residential settlement patterns with cellular automata. J Environ Manage 37:147–160
Diao M (2018) Does growth follow the rail? The potential impact of high-speed rail on the economic geography of China. Transp Res Part A Policy Pract 113:279–290
Dobson A, Richmond E, Simmonds D (2009) Design and use of the new greater manchester land-use/transport interaction model (GM-SPM2). European Transport Conference, Netherlands, pp 411–426
Dorogovtsev SN, Mendes JFF (2004) The shortest path to complex networks. Physics 71:47–53
DSC/ME&P (1999) Review of land-use/transport interaction models. In: Reports to the standing advisory committee on trunk road assessment, pp 12–13
Duan Y, Lu F (2013) Structural robustness of city road networks based on community. Comput Environ Urban Syst 41:75–87
Echenique M (1994) Urban and regional studies at the Martin centre: its origins, its present, its future. Environ Plann B Plann Des 21(5):517–533
Echenique MH, Flowerdew AD, Hunt JD et al (1990) The MEPLAN models of Bilbao, Leeds and Dortmund. Transp Rev 10(4):309–322
Echenique M, Hargreaves AJ, Mitchell G et al (2012) Growing cities sustainably. J Am Plan Assoc 78(2):121–137
Elvidge CD, Baugh KE, Kihn EA et al (1997a) Mapping city lights with nighttime data from the DMSP operational line-scan system. Remote Sens 63(6):727–734
Elvidge CD, Baugh KE, Kihn EA et al (1997b) Relation between satellite observed visible-near infrared emissions, population, economic activity and electric power consumption. Int J Remote Sens 18(6):1373–1379
Elvidge CD, Ziskin D, Baugh K et al (2009) A fifteen-year record of global natural gas flaring derived from satellite data. Energies 2(3):595–622
Feldman O, Davis J, Richmond E et al (2009) An integrated system of transport and land-use models for Auckland and its application. Australasian Transport Research Forum, Auckland, pp 1–5
Feng F, Lei W (2013) Robustness measure of China's railway network topology using relative entropy. In: Discrete dynamics in nature society
Ghosh T, Powell LR, Elvidge DC et al (2010) Shedding light on the global distribution of economic activity. Open Geograph J 3(1):147–160
Givoni M (2006) Development and impact of the modern high-speed train: a review. Transp Rev 26(5):593–611
Golledge R, Stimson RJ (1997) Spatial behavior: a geographic perspective. The Guilford Press, New York
Gonzalez-Navarro M, Turner MA (2018) Subways and urban growth: evidence from earth. J Urban Econ 108:85–106
Gordon M, Anthony H, Anil N et al (2011) Land use, transport, and carbon futures: the impact of spatial form strategies in three UK urban regions. Environ Plann A 43:2143–2163
Government of Ireland (1995) Dublin transportation initiative technical volume I: transport modelling in phase 2. Stationary Office, Dublin
Guidotti R, Gardoni P, Chen Y (2017) Network reliability analysis with link and nodal weights and auxiliary nodes. Struct Saf 65:12–26
Hagerstrand T (1967) Innovation diffusion as a spatial process. University of Chicago Press
Han L, He Z, Tang J et al (2003) A study on dynamic simulation of urban growth and land value increment based on CA model. Geogr Geo-Inf Sci 02:32–35 (in Chinese)
Hansen WG (1959) How accessibility shapes land use. J Am Inst Plann 25:73–76
He D (2009) Complex systems and complex networks. Higher Education Press, Beijing (in Chinese)
He T, Huang L (2020) Impact of high-speed rail network on regional economic coordinated development in Hunan province based on empirical analysis. Scientia Geogr Sci 40(09):1439–1449 (in Chinese)
Holme P, Kim BJ, Yoon CN et al (2002) Attack vulnerability of complex networks. Phys Rev E 65:056109

Horowitz AJ (2004) Lowry-type land use models. In: Hensher DA, Button KJ, Haynes KE, et al (eds) Handbook of transport geography and spatial systems: handbooks in transport, vol 5. Elsevier Science, Oxford, pp 167–183
Hossain M, Alam S, Rees T et al (2013) Australian airport network robustness analysis: a complex network approach. In: Australasian transport research forum (ATRF), 36th
Hu H (1935) The distribution of the population in China—With a statistical table and density map. Acta Geogr Sin 02:33–74 (in Chinese)
Hu X, Huang J, Shi F (2019) Circuity in China's high-speed-rail network. J Transp Geogr 80:1–13
Huang C, Han Q (2019) Does the high-speed rail line have an "Agglomeration Shadow" effect on urban economic activities?—Evidence from satellite lights of cities around the Beijing-Shanghai high-speed railway. Shanghai J Econ 11:46–58 (in Chinese)
Iacono M, Levinson D, El-Geneidy A (2011) Models of transportation and land use change: a guide to the territory. Work Pap 22(4):323–340
Jeswani R, Kulshrestha A, Gupta PK et al (2017) Evaluation of the consistency of DMSP/OLS and SNPP/VIIRS night-time light datasets
Jiang Y (2011) Research on railway network reliability based on complex network theory. Southwest Jiaotong University, Chengdu (in Chinese)
Jiao J, Wang J, Jin F et al (2016) Impact of high-speed rail on inter-city network based on the passenger train network in China, 2003–2013. Acta Geogr Sin 71(2):265–280 (in Chinese)
Kasthurirathna D, Piraveenan M, Thedchanamoorthy G (2013) Network robustness and topological characteristics in scale-free networks. City
Ke X, Bian F (2010) Land-use change ca model based on C5.0 decision tree. Res Environ Yangtze Basin 19(04):403–408 (in Chinese)
Landis J (2001) CUF, CUF II and CURBA: a family of spatially explicit urban growth and land-use policy simulation models. Plan Support Syst—Integr Geogr Inf Syst Mod Visual Tools, pp 157–200
Lei Y, Qian X (2015) Robustness analysis of high-speed railway passenger dedicated line transportation network. J Lanzhou Jiaotong Univer 34:75–80 (in Chinese)
Li X (2007) Geographic simulation system:Cellular automata and multi-agent systems. Science Press, Beijing (in Chinese)
Li X, Yeh AGO (2002) Neural-network-based cellular automata for simulating multiple land use changes using GIS. Int J Geogr Inf Sci 16(4):323–343
Li W, Cai X (2004) Statistical analysis of airport network of China. Phys Rev E 69:046106
Li X, Yeh AGO (2004) Data mining of cellular automata's transition rules. Int J Geogr Inf Sci 18(8):723–744
Li X, Liu X (2006) Case-based cellular automaton for simulating urban development in a large complex region. Acta Geographica Sinica 20(10):1109–1136
Li X, Liu X (2007) Urban evolution simulation based on case-based reasoning and cellular automaton. Acta Geogr Sin 10:1097–1109 (in Chinese)
Li S, Li X, Liu X (2013) Multi-scenario simulations on the interactions of jobs-housing based on agent-based model. Acta Geogr Sin 68(10):1389–1400 (in Chinese)
Li X, Zheng X, Yuan T (2018) Knowledge mapping of research results on DMSP/OLS nighttime light data. J Geo-Inf Sci 20(3):351–359 (in Chinese)
Liang J, Chu B (2005) Urban sprawl and spatially Interdependent development of Beijing. City Plan Rev 6:9–14 (in Chinese)
Lin Y (2017) Travel costs and urban specialization patterns: evidence from China's high-speed railway system. J Urban Econ 98:98–123
Liu J, Yang H (2009) China fights against statistical corruption. Science 325(5941):675–676
Liu Y, Li Y (2017) High-speed rails and city economic growth in China. Finan Res 11:18–33 (in Chinese)
Liu JG, Wang ZT, Dang YZ (2006) Optimization of scale-free network for random failures. Mod Phys Lett B

Liu X, Sheng P, Ma X et al (2018) Temporal and spatial distribution and health risks of vehicle pollution in high-populated urban areas. J Transp Inf Saf 36(1):119–128 (in Chinese)
Long Y, Jin X, Li M et al (2014) A constrained cellular automata model for reconstructing historical arable land in Jiangsu province. Geograph Res 33(12):2239–2250 (in Chinese)
Lowry IS (1964) A model of metropolis RM-4035-RC. Rand Corp, Santa Monica CA
Lu F, Liu K, Duan Y et al (2018) Modeling the heterogeneous traffic correlations in urban road systems using traffic-enhanced community detection approach. Physica A 501:227–237
Luo P, Du Q, He S (2003) The experiment study on spatio-temporal simulation of urbanizing regions based on CA integrated with population density model. Sci Surv Mapp 04:18–21+2 (in Chinese)
Lv Q, Liu H, Wang J et al (2020) Multiscale analysis on spatiotemporal dynamics of energy consumption CO_2 emissions in China: utilizing the integrated of DMSP/OLS and NPP/VIIRS nighttime light datasets. Sci Total Environ 703:134394
Mackett RL (1983) The leeds integrated land-use transport model (LILT). Supplementary Report, TRRL, Crowthorne, pp 805–816
McFadden D (1978) Modelling the choice of residential location. In: Karlquist A et al (eds) Spatial interaction theory and residential location. North Holland, Amsterdam, pp 75–96
Miller EJ, Kriger DS, Hunt JD, Badoe DA (1998) Integrated urban models for simulation of transit and land-use policies. In: Final Report, TCRP project H-12. Joint Program of Transportation, University of Toronto, Toronto
Murray AT, Grubesic TH (2007) Critical infrastructure: reliability and vulnerability
Newman MEJ (2003) The structure and function of complex networks. SIAM Rev 45:167–256
Niu F, Xin Z (2021) Spillover effect of China's high-speed railway stations and its spatial differentiation:An empirical study based on night-time light datasets. Geograph Res 40(10):2796–2807 (in Chinese)
Niu F, Liu W, Song T (2014) A theory implementation and applications of LUTI model. Prod Res 138(4):31–35, 118 (in Chinese)
Niu F, Wang F, Chen M (2019) Modelling urban spatial impacts of land-use/transport policies. J Geog Sci 29(2):197–212
Oryani K, Greiner URS, Britton H (1997) Review of land use models: theory and application. In: Sixth TRB conference on the application of transportation planning methods, pp 80–91
Peng P, Yang Y, Cheng S et al (2019) Hub-and-spoke structure: characterizing the global crude oil transport network with mass vessel trajectories. Energy 168:966–974
Phipps MJ (1992) From local to global: the lesson of cellular automata. Routledge, Chapman and Hall, New York
Pierluigi C, Angel Ibeans I, Luigi DO, Ruben C (2013) LUTI model for the metropolitan area of santander. Urban Plan Dev 139(3):153–165
Pol PM (2003) The economic impact of the high-speed train on urban regions
Qin Y (2016) 'No county left behind?' The distributional impact of high-speed rail upgrades in China. J Econ Geogr 17(3):489–520
Qiu W, Li B, Zhao X et al (2019) Research on layout optimization of residential care facilities in Xuzhou based on GIS. Geomat Spatial Inf Technol 42(08):54–58 (in Chinese)
Reilly WJ (1959) Methods for the study of retail relationships. Bureau of Business Research, Texas
Sallan JM, Lordan O, Simo P et al (2014) Robustness of the air transport network. Transp Res Part E. Logist Transp Rev 68:155–163
Sen P, Dasgupta S, Chatterjee A et al (2003) Small-world properties of the Indian railway network. Phys Rev E 67:036106
Shao S, Tian Z, Yang L (2017) High speed rail and urban service industry agglomeration: evidence from China's Yangtze River Delta region. J Transp Geogr 64:174–183
Shen T, Zhang H, Li X et al (2009) Research on real estate development simulation based on GIS and ABM. Acta Sci Nat Univ Pekin 45(4):653–662 (in Chinese)
Shi Z, Shao J, Pu Z (2018) Effects of quality improvements in transportation infrastructure on total factor productivity growth:Evidence of Chinese railway speed acceleration. World Econ 41(6):127–151 (in Chinese)

Simmonds D, Feldman O (2011) Alternative approaches to spatial modelling. Res Transp Econ 31(1):2–11

Stanislaw I, Krzysztof M (2017) Utilization of cellular automata for analysis of the efficiency of urban freight transport measures based on loading/unloading bays example. Transp Res Procedia 25:1021–1035

Sun Z (1999) Spatial complexity analysis and a geographical model based on cellular automata (GEOCA). Geo-Inf Sci 02:32–37 (in Chinese)

Sun X, Wu Y, Feng X et al (2019) Structure characteristics and robustness analysis of multi-layer network of high speed railway and ordinary railway. J Univer Electron Sci Technol China 048:315–320 (in Chinese)

Thomas RM, Huggett RJ (1980) Modelling in geography: a mathematical approach. Barnes & Noble Books, Totowa, N J

Thunen V, Heinrich J (1826) The isolated state. Perthes, Hamburg

Tian D (2019) A review of studies on the combination of complexity science and urban space research. Urban Stud 26(4):25–30 (in Chinese)

Tobler WR (1970) A computer movie simulating urban growth in the Detroit region. Econ Geogr 46(sup1):234–240

Torrens PM (2000) How land-use transportation models work. Centre for Advanced Spatial Analysis, London

Ureña JM, Menerault P, Garmendia M (2009) The high-speed rail challenge for big intermediate cities: a national, regional and local perspective. Cities 26(5):266–279

Waddell P, Ulfarsson GF (2004) Introduction to urban simulation: design and development of operational models. In: Stopher B, Kingsley H (eds) Handbook in transport, Volume 5: transport geography and spatial systems. Pergamon Press, Oxford, pp 203–236

Wang F (1998) Urban population distribution with various road networks: a simulation approach. Environ Plan b: Urban Anal City Sci 20:265–278

Wang J (2001) Method and practice of integrated planning of overseas land use and transportation. Urban Plan Int 1:5–9 (in Chinese)

Wang J, Jing Y (2017) Comparison of spatial structure and organization mode of inter-city networks from the perspective of railway and air passenger flow. Acta Geogr Sin 72(08):1508–1519 (in Chinese)

Wang Y, Ni P (2016) Economic growth spillover and spatial optimization of high-speed railway. China Ind Econ 2:21–36 (in Chinese)

Wang J, Jin F (2020) China's air passenger transport: an analysis of recent trends. Eurasian Geogr Econ 48:469–480

Wang J, Mo H, Wang F (2014) Evolution of air transport network of China 1930–2012. J Trans Geogr 40

Wang Z, Li M, Jiang J (2020) The impact of traffic accessibility on urban economic growth—spatial econometric analysis of DMSP/OLS nighttime satellite lighting data based on 283 cities. Issues Chin Econ Dev 5 (in Chinese)

Wegener M (2004) Overview of land-use transport models. In: Hensher DA, Button K (eds) Chapter 9 in transport geography and spatial systems. Handbook 5 of the handbook in transport. Pergamon/Elsevier Science, Kidlington, UK, pp 127–146

White R, Engelen G (1993) Cellular automata and fractal urban form: a cellular modelling approach to the evolution of urban land-use patterns. Environ Plan A 25(8):1175–1199

White R, Engelen G, Uljee I (1997) The use of constrained cellular automata for high-resolution modeling of urban land-use dynamics. Environ Plan B 24:323–344

Willianms IN (1994) A model of London and the South East. Environ-Ment Plan B 21(5):535–554

Wolfram S (1984) Cellular automata as models of complexity. Nature 311:419–424

Wu FL (1998) SimLand: a prototype to simulate land conversion through the integrated GIS and CA with AHP-derived transition rules. Int J Geogr Inf Sci 12(1):63–82

Wu F (2002) Calibration of stochastic cellular automata: the application to rural-urban land conversions. Int J Geogr Inf Sci 16(8):795–818

Wu F, Webster CJ (1998) Simulation of land development through the integration of cellular automata and multicriteria evaluation. Environ Plan B 25:103–126
Wu FL, Webster CJ (2000) Simulating artificial cities in a GIS environment: urban growth under alternative regulation regimes. Int J Geogr Inf Sci 14(7):625–648
Xu K, Chen F, Liu X (2015a) The truth of China economic growth:Evidence from global night-time light data. Econ Res J 50(9):17–29, 57 (in Chinese)
Xu F, Zhu J, Miao J (2015b) Robustness study of the air-rail composite network based on complex networks. Complex Syst Complex Sci 12 (in Chinese)
Yang W, Liang J (1997) Advanced economic geography. Peking University Press, Beijing, pp 379–387 (in Chinese)
Yang J, Xie P, Xi J et al (2015) LUCC simulation based on the cellular automata simulation:A case study of Dalian economic and technological development zone. Acta Geogr Sin 70(03):461–475 (in Chinese)
Yang J, Zhang Y, Ge Q et al (2016) Irregular neighborhood cellular automata land-use simulation based on the GA-MCE algorithm. Geogr Res 35(7):1288–1300 (in Chinese)
Yu K, You H, Xu L et al (2012) A minimum cumulative resistance (MCR) analysis approach. Geogr Res 31(7):1173–1184 (in Chinese)
Yu L, Zhao J (2019) Disentangling treatment effect, spatial spillover effect and quantile effect:An empirical study of Beijing-Shanghai high speed rail. World Geogr Res 28(1):47–57 (in Chinese)
Zhang J (2017) High-speed rail construction and county economic development: The research of satellite light data. Economics (quarterly) 16(04):1533–1562 (in Chinese)
Zhang Y, Li X, Liu X et al (2013) Urban expansion simulation by coupling remote sensing observations and cellular automata. Natl Remote Sens Bull 17(04):872–886 (in Chinese)
Zhang J, Cao XB, Du WB et al (2010) Evolution of Chinese airport network. Physica A Stat Mech Appl 389:3922–3931
Zhao J, Ji G, Yue Y et al (2019) Spatio-temporal dynamics of urban residential CO_2 emissions and their driving forces in china using the integrated two nighttime light datasets. Appl Energy 235:612–624
Zheng S, Kahn ME (2013) China's bullet trains facilitate market integration and mitigate the cost of megacity growth. Proc Natl Acad Sci USA 110(14):E1248-1253
Zheng L, Long F, Chang Z et al (2019) Ghost town or city of hope? The spatial spillover effects of high-speed railway stations in China. Transp Policy 81:230–241
Zhou Y, Chen Y (2003) City and urban geography, vol 51. People's Education Press, Beijing, p 41. (in Chinese)
Zhou S, Yan X (2005) The study on the integrated traffic demand–land use models in western countries. Urban Transp China 3(3):64–68 (in Chinese)
Zhou S, Li H (2012) The research summary of land use with the model of cellular automata. Geospatial Inf World 10(5):6–10, 13 (in Chinese)
Zhou C, Sun Z, Xie Y (1999) Research on geographic cellular automata. Science Press, Beijing, pp 78–79 (in Chinese)
Zhou B, Dai T, Liang J et al (2011) A simulation research of nonlinear lowry model based on genetic algorithm. Acta Sci Nat Univ Pekin 47(6):1097–1104 (in Chinese)
Zhou X, Zhang F, Zhou W et al (2012) Evaluating complex network functional robustness by node efficiency. Acta Physica Sinica 19:1–7 (in Chinese)
Zhou B, Dai T, Liang J et al (2013) Simulation of urban spatial structure in Beijing based on the Lowry model. Acta Geogr Sin 68(4):491–505 (in Chinese)
Zipf GK (1946) The PIP2/D Hypothesis: on the intercity movement of persons. Am Sociol Rev 12:677–686
Zondag B, de Bok M, Geurs KT et al (2015) Accessibility modeling and evaluation: the TIGRIS XL land-use and transport interaction model for the Netherlands. Comput Environ Urban Syst 49:115–125

Chapter 4
An Activity-Based LUTI Model—The ActSim Model

This book focuses on elucidating the principles, architecture, and applications of Land Use-Transport Interaction (LUTI) models. Previous chapters have described the principles and architecture of LUTI models. This chapter and subsequent ones will introduce the construction process of specific LUTI models, which are utilized to simulate the spatial distribution patterns of various urban activities, referred to as the ActSim (Activity distribution simulation) model. This chapter will present the architecture of ActSim, while the following chapters will detail the implementation of each model module. The ActSim is constructed using an "activity"-based approach, which evaluates the locational characteristics for different "activities" (Niu 2017; Niu et al. 2015; Zhao and Wan 2020). Here, "activities" refer to various socio-economic activities in the city. Urban activities require "space," generally measured as indoor space in terms of area (Simmonds and Feldman 2011). The "activity"-based modeling technique initially demands a scientific classification of urban activities, where similar activities share comparable locational distribution characteristics, further simulating the spatial patterns of various activities.

The ActSim model is designed to simulate urban activity distribution patterns, thus necessitating the division of urban space into units, considered homogeneous within each unit without further distinction. The size of spatial units is determined by application requirements, referred to as zones.

4.1 Classification of Urban Activities

The LUTI model, emphasized in this book, predicts the location of various urban activities, hence, the principle of activity classification is that similar "activities" have similar locational characteristics. To facilitate the model construction process, urban activities are preliminarily divided into two main categories: households and businesses. Households are where city residents live, and businesses are the places or basic

F. Niu, *Urban Spatial Evolution Simulation*, Urban Sustainability,
https://doi.org/10.1007/978-981-97-3481-8_4

units where residents engage in various economic activities (Xiao 2019), including companies, hospitals, schools, research institutions, and all other work units that can provide jobs. Household location determines the distribution of the residential population, while business location determines the distribution of economic activities. Therefore, predicting the distribution of urban population and economic activities can be categorized into predicting the distribution of households and businesses.

The structure of household members determines their locational preference. Considering data availability, based on the sixth (2010) and seventh (2020) national census data, Beijing households are divided into four categories: single-generation, two-generation, three-generation, and four-generation or more households, calculating the average member structure of each type of household, namely the number of retirees, working population, students, and work-age but unemployed population.

Lowry (1964) classified urban economic activities into two categories based on the service objects: basic activities and non-basic activities (Lowry 1964). Basic activities refer to those that provide services or products to the outside of the city, bringing income to the city, the foundation for the city's existence; while non-basic activities aim to meet the daily production and living needs within the city, serving the city's own operation, which can further be subdivided into activities derived from basic production and those meeting the residents' normal living needs. Lowry believed the location of basic activities is unaffected by the distribution of other activities. However, Beijing, as a megacity with large-scale economic activities, makes it difficult to distinguish basic activities. It is believed that the distribution of all types of activities is influenced by other activities to varying degrees. According to the characteristics of economic activities, businesses are divided into employment, education, medical, and consumption categories. Different businesses have different locational preferences due to their market or service objects. The scale of economic activities can be intuitively represented by the number of people engaged in activities, such as the scale of medical activities measured by the number of patients, and the scale of consumer activities by the number of consumers, but a more operable approach is to evaluate the scale of economic activities by the number of employees in various businesses, such as the number of employees in the medical field representing the scale of medical activities. This book uses "number of employees" to represent the scale of economic activities.

4.2 Prediction of Urban Activity Scale

The scale of various urban activities is constantly changing. To predict their spatial distribution patterns, it is first necessary to determine their total scale. For households, during the city development process, household structures continuously change, leading to changes in the number of households, consumption patterns, and locational needs, thereby affecting the household distribution pattern. The household transition process involves the mutual evolution among various types of households, such as a single-person household becoming a two-person household, a two-person household possibly becoming a three-person household or two single-person households, etc., requiring the establishment of a household transition model to simulate this process. The complexity of the transition model depends on the number of household types. A common method for predicting household scale is to determine future scale based on past changes in household scale, a trend extrapolation method.

Economic activities are completed in businesses, and ActSim calculates the scale of economic activities using business scale. Here, "business scale" refers to the total number of employees in a certain industry, not the number of businesses. Changes in the scale of economic activities manifest as increases or decreases in the total number of employees, which can be determined based on historical change patterns combined with relevant policies to determine the trend of changes in economic activity scale. ActSim aims to predict the overall distribution of urban economic activities, not a specific business, as the location and number of employees of a specific business might not change.

Changes in the scale of urban households and economic activities can serve as input variables, i.e., policy scenarios, by setting the growth rate or number of various households or economic activities to simulate their spatial distribution patterns.

4.3 Factors Influencing the Location of Urban Households and Businesses (Economic Activities)

The ActSim model is designed to predict the distribution patterns of households and businesses. For this purpose, it is first necessary to determine the factors that influence the location of households and businesses. By studying the process of residential location choice, scholars have identified influencing factors as either objective or subjective. Objective factors include housing prices, the accessibility of the housing location, and environmental conditions, while subjective factors include residents' social, economic, and cultural characteristics, as well as their preferences for different locations (Niu and Wang 2018; Zhang 2001). The location choice of businesses is influenced by a variety of factors, resulting from a balance between location costs and institutional environments, industrial development foundations, spatial accessibility, locational conditions, and national land spatial planning policies (Li 2016). In real life, there are many factors influencing the location choices of residents and businesses (Alonso 1964), and it is difficult to exhaustively list them all. The ActSim model mainly considers the major influencing factors, including housing supply, transportation accessibility, and location costs. Additionally, from an implementation perspective, the model can include interfaces to allow input of various variables, enhancing the flexibility of the simulation system.

4.3.1 Housing Supply

Governments control the distribution pattern and use of urban housing, thus influencing the distribution pattern of urban socio-economic activities. Residential housing, as the living space for urban families, is a crucial factor affecting family distribution. Urban housing development is an important means for governments to regulate the distribution pattern of urban family populations, so predicting the distribution of urban housing is necessary before forecasting family population distribution. Similarly, commercial properties (business premises) provide venues for economic activities and are important factors affecting the distribution pattern of businesses (economic activities). Therefore, predicting business distribution requires first determining the distribution pattern of commercial properties.

The distribution pattern of urban housing is determined by the existing stock, demolition, and new development. The new housing area is the sum of the existing stock plus the newly developed area, minus the demolished area, where the demolished area is usually minimal, especially in suburban areas. Thus, the primary task in predicting the urban housing distribution pattern is to forecast the additional area in each zone. Urban new housing development results from the negotiation between policies and developers, where developers aim for maximum profit but are constrained by government planning. Therefore, predicting the additional housing area requires considering both developer and government actions. ActSim simulates the development area that developers pursue for maximum profit under government restrictions. See subsequent chapters for details on predicting urban land use development.

4.3.2 Transportation Accessibility

The convenience of transportation is a crucial factor affecting the location choices of urban families and businesses. The indicator representing the convenience of transportation is usually referred to as transportation accessibility. The "activity"-based LUTI model (ActSim) evaluates transportation accessibility separately for different "activities" (Niu et al. 2016). For families, the accessibility of a zone should reflect the convenience with which family members can engage in various economic activities from that zone. For example, by dividing urban economic activities into employment, education, medical, and consumption, the corresponding accessibility evaluations are employment accessibility, educational accessibility, medical accessibility, and consumption accessibility. Different families, due to their member structures, prioritize different types of accessibility. For instance, two-person households, mostly working-age population, would prioritize locations with high employment accessibility, while families with school-aged children would choose locations with high educational accessibility. Therefore, it is necessary to synthesize various types of accessibility based on family member structures to evaluate household accessibility.

For businesses, as the endpoints of trips, the ease with which they can be reached needs to be considered. Therefore, this book introduces the concepts of origin accessibility and destination accessibility. Origin accessibility, for the starting points of trips, represents the convenience of engaging in various economic activities from a certain point. The aforementioned types of accessibility for families are all examples of origin accessibility. Destination accessibility represents the convenience of reaching a place as a destination. The location choice of businesses is influenced by various types of destination accessibility. For example, educational businesses consider the convenience for school-aged population to reach them, focusing on the distribution of school-aged population, while medical institutions prioritize the spatial distribution of the elderly. For methods of evaluating transportation accessibility, see subsequent chapters.

4.3.3 Location Costs (Rent)

Location cost refers to the economic cost of living or operating in a certain location for families or businesses, mainly determined by rent (or housing prices), and can consider other expenses, such as property management, water, and electricity fees, where data support. For families, a household typically allocates its income between housing and other consumption to maximize utility (Alonso 1964; Zhao and Wan 2020). Families with different incomes have different proportions (or tendencies) of income allocated to various consumption items. When income allocation is fixed, the same living cost corresponds to a smaller living area as rent increases. Therefore, due to different rents, the consumption utility of different plots also varies. The consumption utility of each zone can be calculated using the Cobb–Douglas equation (Cobb and Douglas 1928). For businesses, which typically aim for profit, the main concern is absolute costs such as rent. For calculations of location costs, see subsequent chapters.

4.3.4 Other Influencing Factors

In addition to the aforementioned factors, the location of urban households or businesses is influenced by various other factors, including environmental quality, population quality, religious culture, etc. Assuming data availability, all such factors can serve as criteria for location evaluation. The impact of each factor on households and businesses varies; thus, when predicting the location of households or businesses, it is necessary to determine the weight of each criterion. Evaluation of each influencing factor requires the establishment of corresponding sub-models.

4.4 Composition of the ActSim Model

ActSim assumes that all urban activities tend towards locations with higher utility, based on a comprehensive evaluation of the above-mentioned locational influencing factors, to predict the distribution patterns of households and businesses. ActSim includes two major sub-models: a transportation model and a location model. The transportation model evaluates urban traffic accessibility based on urban transportation costs and the distribution of urban activities. The location model predicts the distribution patterns of urban households and businesses based on housing distribution, location costs, and transportation accessibility, among other indicators. The composition of the ActSim model is shown in Fig. 4.1.

The input to the ActSim model is the current distribution of urban activities, including residential population and business (employment) distribution, rent, housing distribution, and the transportation network. The output is the distribution patterns of the residential population and businesses. The transportation model calculates urban traffic accessibility, which serves as an input to the location model, used to calculate the locations of households and businesses.

The specific implementation of the above modules will be detailed in subsequent Chap. 5 Through 8. Chap. 5 details the urban housing development prediction model; Chap. 6 explains the method for evaluating urban location costs, including the evaluation methods for the location costs of households and businesses; Chap. 7 elucidates the method for evaluating urban transportation accessibility, including the evaluation methods for origin accessibility (OA) and destination accessibility (DA); Chap. 8 clarifies the implementation of the location model, which synthesizes all the

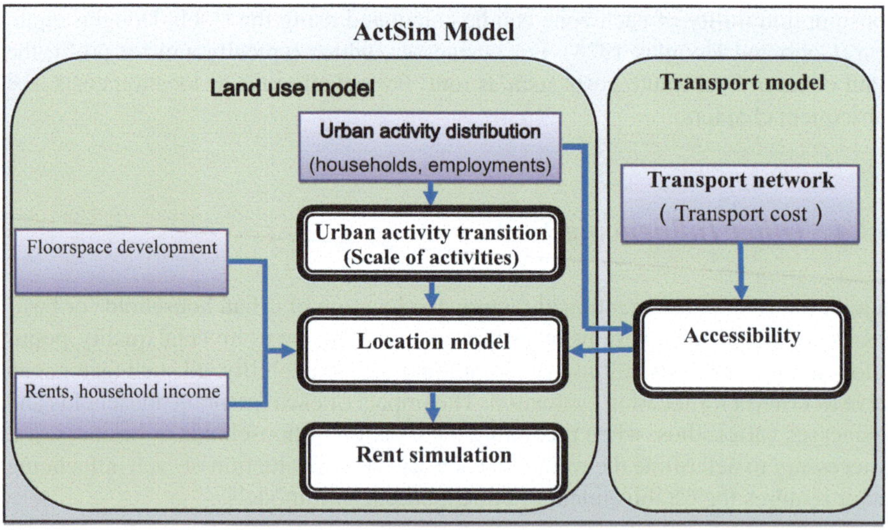

Fig. 4.1 Composition of the ActSim model

sub-models to predict the locations of urban households and businesses (economic activities). Chapters 5–8 are closely related, detailing the specific implementation of each module of the ActSim model, forming the core content of this book.

4.5 Chapter Summary

This book is dedicated to establishing an "activity"-based LUTI model, named ActSim, for simulating and predicting the spatial distribution patterns of urban activities. This chapter focuses on clarifying the architecture of the ActSim model, with subsequent chapters detailing the specific implementation of each sub-module. The components of ActSim include sub-models for predicting the scale of urban activities, housing development prediction (land use), evaluation of location costs, the location model, and the transportation model (evaluation of transportation accessibility). Additionally, changes in the distribution of urban family populations and businesses will lead to changes in rent, which will, in turn, be used to predict the distribution patterns of households and businesses again. Therefore, the model algorithm embeds a rent model for calculating rent.

The urban activity scale prediction sub-model determines the total scale of various households and businesses; the housing development prediction sub-model forecasts the distribution pattern of residential and commercial property development, i.e., urban land use development; the location cost evaluation sub-model determines the economic cost for a family or business to choose a particular location; the location model determines the locations, i.e., the distribution patterns, of various households or businesses; the transportation sub-model evaluates the accessibility of different urban zones, representing their transportation advantages. The specific implementation of the model depends on application requirements and data availability.

References

Alonso W (1964) Location and land use: toward a general theory of land rent. Econ Geogr 42(3):11–26

Cobb CW, Douglas PH (1928) A theory of production. Am Econ Rev 18(Supplement):139–165

Li Y (2016) Discussion on the relationship between production factor costs, enterprise location choices, and industrial spatial distribution. Bus Econ Res 20:108–110 (in Chinese)

Lowry IS (1964) A model of metropoils. Rand Crop, Santa Monica

Niu F (2017) Overview of urban land-use/transport interaction model: Origin, techniques and future. Sci Geogr Sin 37(1):46–55 (in Chinese)

Niu F, Liu W, Feng J (2016) Modeling urban housing price: The perspective of household activity demand. Acta Geogr Sin 71(10):1731–1740 (in Chinese)

Niu F, Wang F (2018) Modelling urban spatial impacts of land-use/transport policies. Acta Geogr Sin 73(2):380–392 (in Chinese)

Niu F, Wang Z, Hu Y et al (2015) A model of urban spatial evolution process based on economic and social activities. Prog Geogr 34(1):30–37 (in Chinese)
Population Census Office of the State Council (2012) China 2010 population census data. China Statistics Press, Beijing, pp 798–799. (in Chinese)
Simmonds D, Feldman O (2011) Alternative approaches to spatial modelling. Res Transp Econ 31(1):2–11
Xiao L (2019) Economic geography. Science Press, Beijing (in Chinese)
Zhang W (2001) An analysis of the factors that influence the urban residential location selection. Prog Geogr 03:267–274 (in Chinese)
Zhao P, Wan J (2020) The key technologies of integrated urban transport-land use model: Theory base and development trends. Sci Geogr Sin 40(1):12–21 (in Chinese)

Chapter 5
Simulation and Prediction of Urban Housing Development

Cities are places where residents engage in socio-economic activities, and urban buildings provide venues for these activities. The government controls the spatial distribution and purpose of urban buildings, thereby regulating the distribution pattern of residents' socio-economic activities. Urban buildings, developed by developers, are indirectly controlled by the government through the regulation of developers' behavior, i.e., urban land use. Urban buildings can be broadly classified into residential and commercial properties, which can be further subdivided into various types. Below, we take residential development as an example to illustrate the simulation and prediction method for housing development. The prediction method for commercial property (enterprise housing) development is similar.

5.1 Simulation of Urban Residential Development

Urban residential development is the result of a game between developers and the government, i.e., the result of both market and policy actions. Developers pursue maximization of market benefits while being restricted by policies (Niu and Wang 2020).

Typically, when the government sells land, it specifies the maximum development area for housing or commercial properties, called "Permitted Development Area" (PD). Developers make decisions based on the permitted development area and profits in recent years to maximize profits within the permitted range. Based on this, this paper constructs a Housing Development Model (Floorspace Development Model, FDM) to predict the distribution of urban housing development by simulating developer behavior. This model considers both market and government policy actions and predicts the urban housing distribution pattern in four steps.

F. Niu, *Urban Spatial Evolution Simulation*, Urban Sustainability,
https://doi.org/10.1007/978-981-97-3481-8_5

First, predict the total unrestricted development area F(U) of urban housing, i.e., the total scale of housing that developers might develop to maximize profits without considering government development permits.

Second, determine the total constrained development area F(C), i.e., revise the predicted development area from the previous step according to the government's permitted development area, ensuring the total development scale does not exceed the permitted development area.

Third, determine the spatial distribution pattern of urban housing development. Based on the profit rate of each urban zone and the permitted development area of each zone, distribute the total development area among the zones, ensuring the development area of each zone does not exceed its permitted development area.

Finally, update the total housing volume of each zone to determine the urban housing distribution pattern. The new development area of housing in each zone, added to the existing housing area, yields the new total housing volume. The construction of new housing inevitably leads to changes in rent, so further predictions of rent changes are needed, which in turn cause changes in population distribution. The prediction of population distribution patterns will be implemented by the location model, detailed in Chap. 8. The specific implementation of the above steps is as follows.

5.1.1 Prediction of Total Unrestricted Housing Development Area

It is assumed that the primary influencing factor in a developer's decision is profit, and they will also consider the existing housing distribution pattern, referencing profits from recent years. Variables representing profit include rent (or house price), demolition costs, construction costs, etc. Under perfect market conditions, rent is the comprehensive manifestation of all these factors and the most critical. Generally, an increase in the rent of a particular zone signifies rising housing demand at that location, leading to further increases in housing development area in the next time period. If year *t* is taken as the base year, the basic form of the prediction function for the total unrestricted development area $F(U)_p$ in the next time period *p* is as shown in Eq. (5.1). Here, t represents the current year, focusing on a specific point in time, while *p* in Eq. (5.1) refers to the entire year of $t + 1$, emphasizing a period.

$$F(U)_p = \alpha_p . F_t . \prod_v \left(\prod_{y=0}^{N} \left(v_{(t-y)} \right)^{\beta} \right) \tag{5.1}$$

In this equation, α_p is a proportional factor, and F_t is the total housing scale of the city at the end of year *t*; $v_{(t-y)}$ is the value of variable *v* in year $(t - y)$, such as rent, demolition costs, etc.; β is the exponential parameter of the variable, which is the adjustment parameter for each variable; y is the lag time, i.e., the number of years affecting the developer's decision, ranging from 0 to N. For example, if $N = 3$, it indicates that the developer's decision considers the situation of the past three years.

5.1.2 Constrained Development Area

The development area under policy constraints is known as the constrained development area, denoted as $F(C)$. To determine the constrained development area, the unrestricted development area $F(U)$ must be compared with the permitted development area. Let $F(P)_{pi}$ be the permitted development area of zone i in the time period p (usually the next year). The sum of the permitted development areas of all zones is the city's total permitted development area $F(P)_p$, as shown in Eq. 5.2.

$$F(P)_p = \sum_i F(P)_{pi} \tag{5.2}$$

If the unrestricted development area $F(U)_p$ does not exceed the city's total permitted area $F(P)_p$, i.e.,

$$F(U)_p \leq F(P)_p \tag{5.3}$$

then policy constraints will not affect the unrestricted development area predicted in the previous step, and the constrained development area equals the unrestricted development area:

$$F(C)_p = F(U)_p \tag{5.4}$$

If the unrestricted development area exceeds the permitted development area, i.e.,

$$F(U)_p > F(P)_p \tag{5.5}$$

then the constrained development area $F(C)_p$ equals the permitted development area:

$$F(C)_p = F(P)_p \tag{5.6}$$

5.1.3 Housing Development Spatial Distribution—Determining Development Area for Each Zone

After predicting the total housing development area (constrained development area) for the city, the next step is to distribute this total development area among each zone, determining the development area for each zone. The distribution of the total development area to each zone needs to consider the profit rate of each zone as well as policy permissions. Assuming that developers prefer zones with higher profits and that the actual development area is proportional to the permitted development area, a model is established to predict the new development area of zone *i* in period *p*, as shown in Eq. (5.7), referencing the discrete choice model (McFadden et al. 1978).

$$F(N)_{pi} = F(C)_p \cdot \frac{F(P)_{pi} \cdot \exp(\gamma_p \cdot r_{ti})}{\sum_i F(P)_{pi} \cdot \exp(\gamma_p \cdot r_{ti})} \tag{5.7}$$

Here, $F(C)_p$ is the total constrained development area for the city in period *p*, $F(P)_{pi}$ is the permitted development area for zone *i*, r_{ti} is the rent for zone *i* in year *t*, and γ_p represents the sensitivity of housing development to profits. When the value of γ_p is large, the spatial allocation of development areas tends to favor zones with higher profits; when the value of γ_p is small, it indicates that housing development is not sensitive to profits.

At this point, the development area for each zone is preliminarily determined. Since each zone has policy restrictions, the model needs to check whether the new development area for each zone exceeds the permitted development area for that zone, i.e., whether Eq. (5.8) holds.

$$F(N)_{pi} \leq F(P)_{pi} \tag{5.8}$$

If the above equation does not hold, meaning the allocated development area for zone i exceeds the permitted area, then the excess portion will be subtracted. The excess portions from each zone are then added up and redistributed according to Eq. (5.7). This process is iterative until the development area for all zones satisfies Eq. (5.8).

After this step, the new development area for each zone is determined. The original housing area of each zone plus the new development area equals the total area. Changes in housing distribution will affect rent and the distribution of the resident population, thereby affecting the housing development pattern for the following year, creating a cyclical process. Prediction of rent and family population distribution requires the use of the ActSim location model, hence, this model operates in conjunction with other sub-models to predict housing distribution for subsequent years. The implementation of the location model is detailed in Chap. 8.

5.2 Case Study: Simulation of Housing Development in Beijing

This section illustrates the application of the FDM model using Beijing as a case study. As the capital of China and a typical megacity, Beijing has experienced rapid population growth and continuous expansion of construction land in recent years. According to the requirements of the Beijing-Tianjin-Hebei coordinated development strategy and the needs of modern urban system construction, Beijing will relocate some economic activities and residential populations, facing significant spatial changes. Therefore, Beijing is chosen as the case area to simulate the spatial distribution of urban housing development using the FDM. As mentioned earlier, since the FDM requires rent data, which is affected by population distribution, the operation of the FDM model needs to be integrated with the location model (predicting population and rent). In other words, the output of the FDM model is only part of the output of the ActSim model.

5.2.1 Data

Beijing's population and economic activities are highly concentrated in the central urban area and the suburbs. Therefore, this study selects the area within the Sixth Ring Road as the research area, with streets (townships) as the research units. The research area includes a total of 239 streets or townships, referred to as zones. The required data include policy permission (representing land use policy), existing building distribution, and rent (excluding demolition costs and other factors for now).

The study collected Beijing's land transaction data from 2009 to 2013 (Beijing Land and Resources Bureau 2014); estimated the housing construction area for each zone based on population data and per capita housing area (National Bureau of Statistics 2010; Beijing Bureau of Statistics 2013) obtained rent data from housing transactions on Fang.com, further processing the data to exclude high or low values (outliers). Since the model is based on market principles, unusually high or low rents or prices are due to special reasons and do not result from normal market laws.

5.2.2 Model Calibration

Model calibration is a crucial step in the simulation work. Although detailed building distribution data for model calibration is not available, rent plays a vital role in the simulation results and is usually affected by the supply–demand relationship, directly linked to the development area. To calibrate the model, each coefficient in the model is initially set to an empirical value, and then the model predicts housing development distribution. Coefficients are adjusted by comparing predicted values with observed

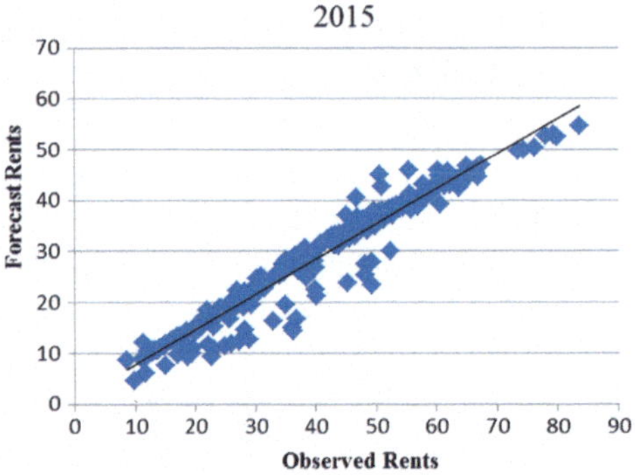

Fig. 5.1 Correlation between the predicted rent values and actual observed values of various zones in Beijing in 2015

values. For instance, if all zone rent predictions are higher than observed values and the model's R-squared is less than 0.50, the coefficients positively correlated with rent are reduced by 0.5%, and vice versa. Other coefficients are adjusted similarly until predicted values closely match actual observed data. Using 2014 as the base year, predictions are made annually. Figure 5.1 shows a comparison of the predicted rent for 2015 with actual observed values, achieving an R-squared of 0.76, indicating a good fit.

5.2.3 Housing Development Simulation

1. **Land Use Policy Scenario Setting**

The policy scenario set in this study is for experimental purposes and does not represent official policy. Assuming that the policy keeps the permitted development area (PD) stable for each spatial unit, extending to future years, the average of the housing development permitted area from 2009 to 2013 (derived from land transaction data) is used as the PD value for years after 2014, as shown in Fig. 5.2. Zones with larger PD values are mainly distributed along the Sixth Ring Road, due to most areas in the city center being developed areas with limited development potential. The spatial distribution of PD aligns with Beijing's planning goals to develop the periphery of the central city.

2. **Prediction of Housing Distribution Pattern for 2020**

Using 2014 as the base year and based on the set policy scenario (PD), the FDM is used to predict housing development distribution for 2015 and subsequent years annually. To predict housing development for 2015, rent data for 2014 is used as input data. Predictions for subsequent years require the previous year's rent data, obtained using the location model. Subsequent chapters will detail the specific implementation of each part of the ActSim model, while the goal of this chapter is to elucidate the functionality and implementation steps of the FDM.

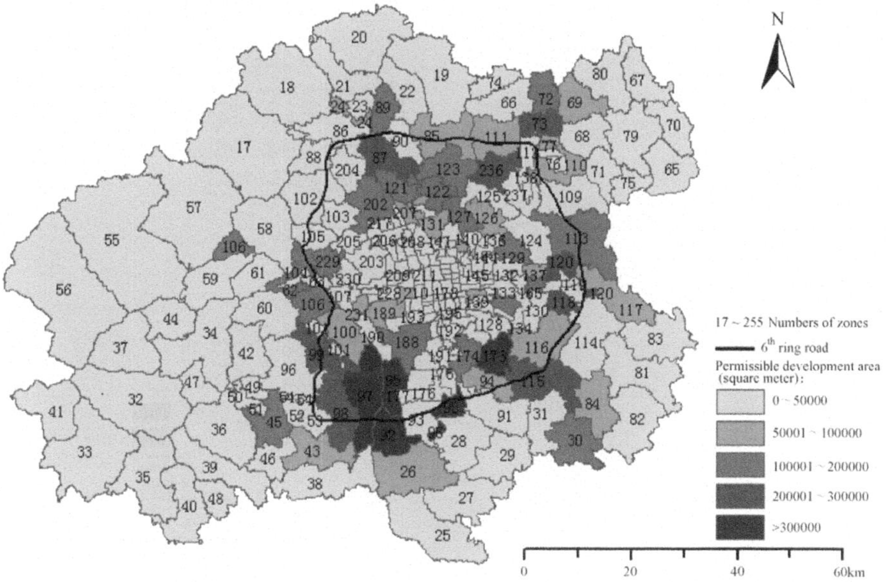

Fig. 5.2 Zonal permissible development distributions in Beijing

The prediction results for the housing distribution pattern in 2020 are shown in Fig. 5.3. The figure indicates that housing in Beijing in 2020 is still predominantly distributed within the Fifth Ring Road. Moreover, transportation routes significantly influence the spatial layout of buildings, with a tendency to develop near major roads and areas with convenient transportation.

Comparing the housing distribution in 2020 with that in 2010, the spatial growth pattern of housing is analyzed. Figure 5.4a and b show the spatial distribution of housing area growth in absolute values and percentages, respectively (classified using natural breaks). Both figures reveal that housing development from 2010 to 2020 mainly occurs outside the Fifth Ring Road, especially between the Fifth and Sixth Ring Roads, while the development intensity in the central urban area is weaker. This is due to the high development costs in the already highly developed central urban area and the government's development policy to relieve the population of the central city. Additionally, the high development intensity naturally arises near the Fifth and Sixth Ring Roads, where transportation accessibility is higher, leading to higher development profits.

Based on the growth rate of housing, zones are classified into five types: high growth, relatively high growth, medium growth, relatively low growth, and low growth. Zones categorized as high growth and relatively high growth include Mapo (73), Liangxiang (98), Beizang (92), Changyang (97), Yizhuang (173), Nanshao (89), Houshayu (236), Wulituo (229), Wangzuo (99), Yongding (106), Niulanshan (72), and Longquan (107), distributed near the administrative centers of Shunyi, Fangshan, Daxing, Changping, Shijingshan, Fengtai, and Mentougou districts. This indicates

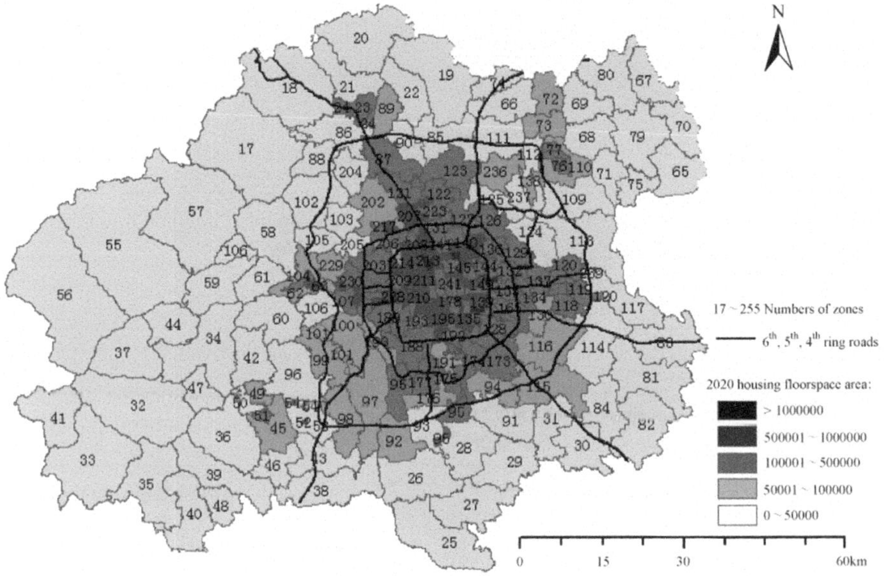

Fig. 5.3. 2020 housing floorspace distribution

that although housing development tends towards the suburbs, it still concentrates near the county towns of each suburban district. The government is guiding Yizhuang (173) to become a secondary center of Beijing, offering many preferential policies for enterprises settling in this zone with a high PD value.

Similarly, the FDM can also be used to predict commercial housing development. To predict commercial housing development, the model's parameters need to be replaced with relevant indicators for commercial housing. Corresponding to the prediction of residential housing development, the four steps for predicting commercial housing development are: first, simulate the development area for maximum profit without policy restrictions; second, calculate the total commercial housing development area under policy restrictions; third, distribute the aforementioned restricted development area among all zones based on profit and policy restrictions to determine the development area for each zone; fourth, add the development area of each zone to the existing stock to obtain the total commercial housing area for that zone.

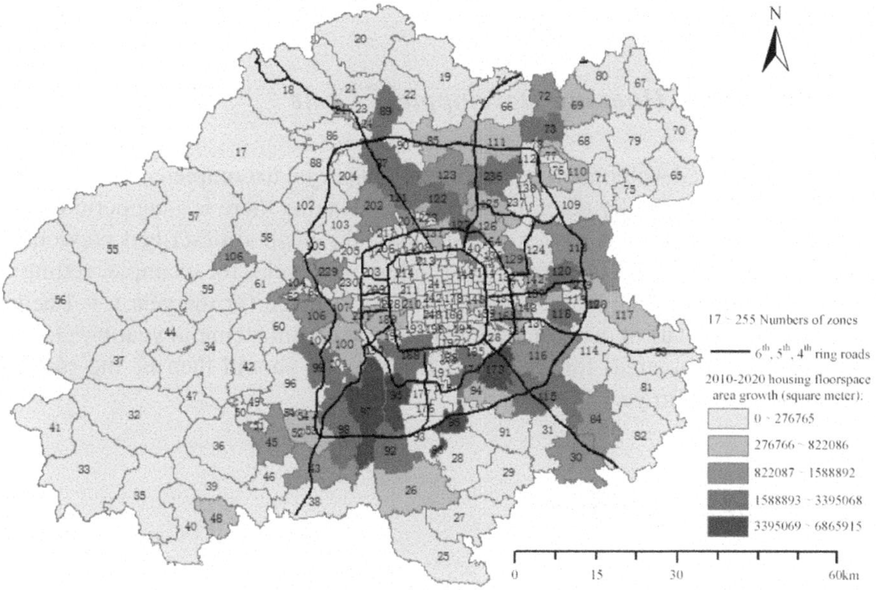

a. Spatial Pattern of Housing Size Growth from 2010 to 2020 (Absolute Value)

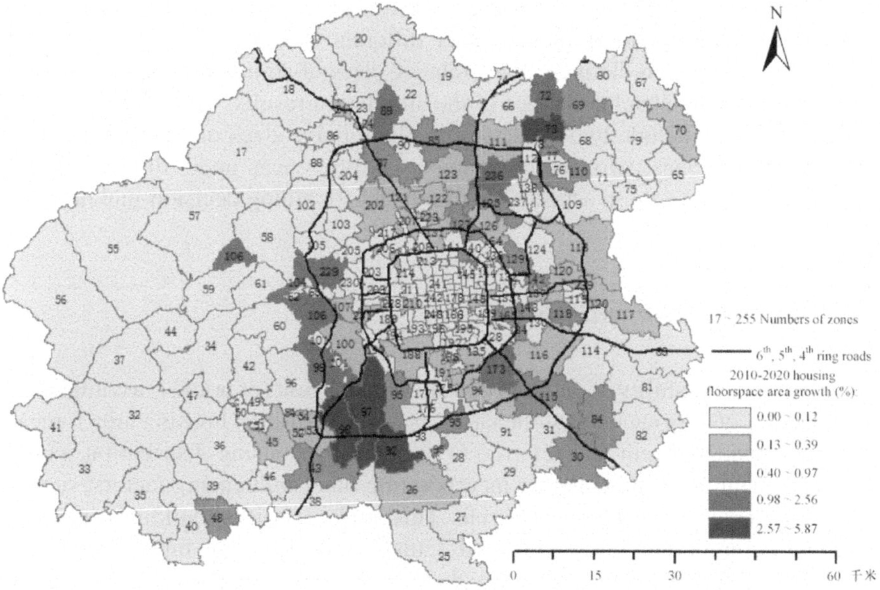

b. Spatial Pattern of Housing Size Growth from 2010 to 2020 (Percentage)

Fig. 5.4 Spatial pattern of housing size changes in Beijing from 2010 to 2020 (in Chinese)

5.3 Discussion

5.3.1 About the Housing Development Cycle

Urban construction development requires a certain period to complete. The development of a building typically takes about 2 years, meaning there is a lag period from the start of development to its actual use. Thus, the method described previously ignores this lag period. Considering the development cycle, for example, setting a 2-year development cycle, the development area predicted for the year t + 1 actually represents the start of construction, with its use commencing in year t + 3. The completed area (new area) for years t + 1 and t + 2 should then be the start of construction area from years t − 1 and t.

5.3.2 Accuracy of FDM

Regarding the use of the model, since urban spatial development is a complex process, it is difficult to predict accurately. FDM is not a model that can precisely predict the future but provides a policy scenario testing tool to examine the spatial distribution of urban housing development under different land use policy scenarios. In terms of the model's predictive values, relative sizes of simulation values are more meaningful than the absolute values of the simulation results. The relative sizes of simulation values include two aspects: one, the relative sizes of simulation values among different zones; two, the comparison of differences under different policy scenarios. We believe these aspects are significant for aiding decision-making.

5.3.3 Model Optimization

Limited by the availability of data and considering that rent (or price) is a comprehensive reflection of various factors, the case study's unrestricted housing area forecast primarily considered rent and referred to the existing housing distribution. Future model optimization could consider incorporating other influencing factors, such as construction costs, demolition costs, etc. This study classified buildings into residential and commercial categories, and actual application can further refine this classification based on data availability and application needs, such as subdividing commercial buildings into service, office, education, etc., to achieve more detailed classification predictions. This article is a methodological study, and the model can be modified and calibrated to different extents for further application in other cities. Through more case applications and comparisons, continuous improvements in variable selection and parameter setting can gradually perfect the model.

5.3.4 Simulation of Urban Land Market

FDM predicts housing development based on profit, while in reality, urban housing development is constrained by land supply. Developers need to secure land before developing properties. Given the same profit, if the land supply is different, the development area of the houses will also differ. Therefore, a land market model is needed to predict the spatial distribution of land supply. The land supply should be determined first before further predicting the housing development area. Of course, FDM will need corresponding improvements to establish a quantitative relationship between land market and construction development area.

5.4 Chapter Summary

Cities are places where residents engage in socio-economic activities, and urban buildings provide venues for these activities. The government controls the spatial distribution and purpose of urban buildings, managing the distribution pattern of urban resident activities, i.e., urban land use. Urban building development is completed by developers, who, while seeking to maximize profits, are constrained by policies. This chapter constructed a housing development model, called FDM, to predict the urban housing distribution pattern. The FDM model predicts urban housing development in four steps. First, it predicts the development area for developers seeking to maximize profits, known as the unrestricted development area; second, it compares the unrestricted development area with policy restrictions to calculate the restricted development area; third, it determines the housing development distribution pattern, i.e., distributing the restricted development area among various zones; lastly, it adds the predicted development area to the existing housing area to obtain the total housing area for each zone.

FDM can also be used to simulate commercial property development, requiring the corresponding variables to be switched to those relevant to commercial properties. The distribution of urban buildings affects the distribution patterns of urban populations and businesses and will be used as input data to predict the distribution patterns of families and businesses.

References

Beijing Bureau of Statistics (2013) Beijing population and employment statistical yearbook. China Statistics Press, Beijing (in Chinese)

Beijing Land and Resources Bureau (2014) Land grant announcement. http://bjtd.bjgtj.gov.cn/default.aspx, bjgtj.gov.cn (in Chinese)

McFadden D (1978) Modelling the choice of residential location. In: Karlquist A et al (eds) Spatial interaction theory and residential location. North Holland, Amsterdam, pp 75–96

National Bureau of Statistics (2010) The sixth national population census data. China Statistics Press, Beijing (in Chinese)
Niu F, Wang F (2020) Modelling urban housing development and a Beijing case study. Geograph Sci 40(1):97–102 (in Chinese)

Chapter 6
Urban Location Cost Evaluation

Location cost, or the cost associated with the locational characteristics of production and living input factors that exhibit significant locational differences, is a crucial determinant of the economic cost for households or businesses in choosing a particular location. This, in turn, influences their site selection and ultimately the distribution pattern of urban population and businesses (economic activities). The collective site selection behaviors of micro-entities drive the evolution of urban spatial structures.

Rent (or housing price) is a major factor influencing location costs. Different types of households or businesses require different types of buildings, leading to varying location costs. For example, households with four generations require larger homes compared to single-person households, resulting in different corresponding rents. Similarly, the building needs of businesses vary; for instance, storage-based businesses have different building requirements compared to office-based businesses, affecting their respective rents. Therefore, the evaluation of location costs also proceeds by the type of economic activity. It's noteworthy that building classification and economic activities do not correspond exactly, as different economic activities may require the same type of building. For instance, both couples and single-person households may opt for the same house type, just as office-based and education-based businesses may use the same building types. This chapter discusses the method for evaluating urban location costs for both households and businesses.

F. Niu, *Urban Spatial Evolution Simulation*, Urban Sustainability,
https://doi.org/10.1007/978-981-97-3481-8_6

6.1 Household Location Cost Evaluation—Consumer Utility

6.1.1 Overview of Household Location Cost Evaluation

The utility derived from housing depends not only on the building itself but also on its spatial location, i.e., residential location. The simulation of housing prices traces back to Thünen's agricultural location theory, which abstractly expresses the law of urban land price distribution based on a series of assumptions (Krugman 1996). Building on this, Alonso, Mills, and Muth proposed the monocentric city model in the 1960s, laying a theoretical foundation for urban spatial simulation. The core of this model analyzes the trade-off behavior between housing consumption and commuting to the city center (Fujita 1989). Furthermore, different types of households have varying preferences for residential locations, leading to spatial differentiation in population distribution (Niu et al. 2016). Most residents freely choose housing that meets their needs on the housing market, based on their financial capacity, personal preferences, and workplace locations. The willingness to pay for different residential locations reflects their preferences or demand characteristics for those locations (Zheng et al. 2005).

Consumer utility, a measure of the degree of psychological satisfaction of consumers, is generally pursued as a principle in consumption activities. Under the constraint of a household budget, rational households always optimally allocate their limited consumption expenditure across various bundles of goods and services. By adjusting the proportion among these bundles, they seek to maximize consumer utility while meeting their needs. Thus, in evaluating household location costs, ActSim combines household income to assess their consumer utility.

6.1.2 Household Consumer Utility Evaluation Method

Household expenditures cover a wide range of goods and services. We categorize household spending into two types: housing consumption and consumption of other goods or services (OGS). The Cobb–Douglas production function is used to evaluate urban consumer utility. It's important to note that transportation costs are not considered in this utility calculation because they are already included in the evaluation of transportation accessibility within the ActSim model.

The Cobb–Douglas production function, widely applied in economics, was developed based on data from the U.S. manufacturing sector from 1899 to 1922 by mathematician Cobb and economist Douglas. They explored the relationship between input and output, introducing this function in 1928 (Cobb and Douglas 1928). This function is applied here to calculate the consumer utility of urban households as shown in Formula (6.1a).

$$U_i = (a_i^H)^{\beta^H} \cdot (a_i^O)^{\beta^O} \tag{6.1a}$$

$$\beta^H + \beta^O = 1 \tag{6.1b}$$

In which, U_i represents the consumer utility of households in zone i; a_i^H is the average housing area for the household; a_i^O is the average expenditure of the household on *OGS* (Other Goods or Services); β^H and β^O represent the propensity of households to allocate their income between housing and OGS, i.e., the proportion, the sum of which equals 1. For households of the same type with similar income levels, their propensity to allocate income between housing and other consumptions is usually similar. When the proportion of income allocation is fixed, due to different rent levels across zones, the same income can afford a smaller living area in areas with high rent, thus reducing the consumer utility (U_i). Conversely, in locations with lower rent, a larger living area is possible, enhancing consumer utility. The consumer utility of a zone is influenced by the household's income level.

6.1.3 Case Application: Evaluation of Household Consumer Utility in Beijing

Spatial Scale: The evaluation of consumer utility in the city is conducted by zone. We use streets (townships) as spatial units. The northern districts of Beijing, Yanqing, Huairou, Miyun, and Pinggu, contain large areas of mountains and rural regions, so they are not divided further; instead, data from their county towns are used.

Data: Evaluating household consumer utility requires information on the income of various types of households and the rent in each zone, which is generally difficult to obtain. For this example, we use the average income of households across the city. Rent data is sourced from Fang.com. The proportion of income allocation for households is determined based on interviews, enabling the calculation of the living area that the same income can afford in each block. When expenditures are fixed, the quantity of other goods consumed by the household is assumed to remain constant.

Based on the above settings, the consumer utility of each zone in Beijing can be calculated, as shown in Fig. 6.1. The figure reveals that areas with high consumer utility are mostly distributed in the outer suburbs, while low utility areas are mainly located in the city center. The distribution pattern of consumer utility generally increases from the city center to the suburbs. This is because locations closer to the city center have higher rents, meaning the same expenditure corresponds to a smaller living area and lower consumer utility.

Urban residents have different incomes due to the industries they work in. Households with different incomes have different proportions of housing expenditure, leading to varying consumer utility in each zone for different households. This also results in diverse location choices for different households. Therefore, to achieve a

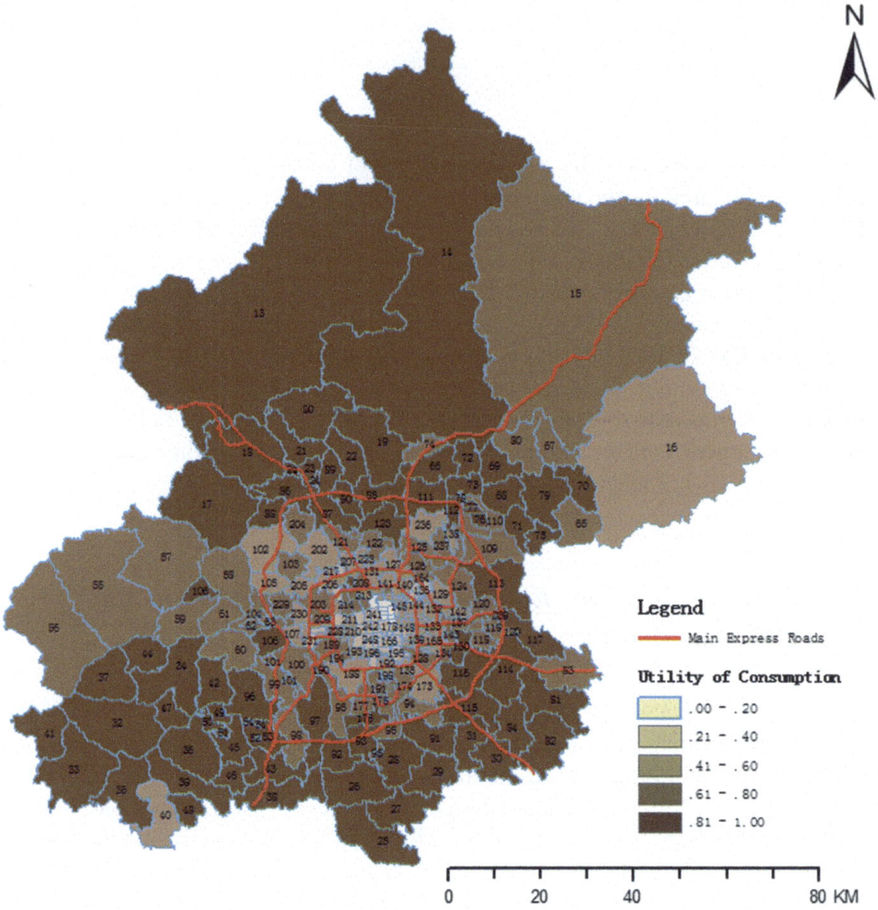

Fig. 6.1 Consumer utility of residents in Beijing

more detailed evaluation of urban consumer utility, it's necessary to divide households into different socio-economic groups based on their income and preferences for residential and other consumptions, and then evaluate the consumer utility of each zone for each group.

Additionally, location costs also include property management fees, heating fees, water and electricity fees, etc. When data is available, these factors can be considered. Typically, under market economy conditions, rent can comprehensively reflect the location cost of each zone.

6.2 Enterprise Location Cost Evaluation

Due to the spatial separation of production activities, there are contact costs within and between enterprises, such as those between headquarters and production factories of multi-location enterprises, and those arising from vertical division of labor between different enterprises. Contacts include information flow (like business negotiations, dialogues) and material flow (like specific products). Economic activities tend to agglomerate due to the existence of contact costs, reducing transportation (material contact cost) and information contact costs, bringing positive externalities to enterprises. Such spatial agglomeration generates urbanization and localization economies (Fan and Li 2011).

As analyzed previously, enterprise costs include production costs and contact costs (or transaction costs), with the former referring to various production input factors (labor, raw materials, machinery, etc.) and rent, and the latter to the aforementioned material and information contact costs. Contact costs and rent together constitute the spatial cost (Shao 2012). When other input costs are fixed, enterprises, driven by profit maximization, seek to minimize spatial costs. Different types of economic activities have varying dependencies on location. Labor-intensive enterprises prefer to be near city centers, while capital-intensive enterprises choose less accessible peripheral areas (Scott 1982).

Practically, we directly use rent to calculate enterprise location costs. However, different enterprises have varying space requirements and density (the same scale of enterprise requires different building areas), so per-unit rent does not accurately represent location costs for all types of enterprises. Accordingly, the ActSim model calculates per capita rent for each type of enterprise. For instance, for educational enterprises, the per capita building area is calculated based on the total floor area and the total number of employees in the education sector in a zone. This figure is then multiplied by the per-unit rent to determine the per capita rent, representing the location cost for educational enterprises.

6.3 Chapter Summary

Location cost refers to the economic cost paid by households or enterprises for choosing a particular location and is a significant factor affecting their site selection. Rent (or housing price) is a major factor influencing location cost. The considerations of urban households and enterprises for location cost differ markedly. Households seek to maximize consumer utility when allocating their limited income to housing or other consumptions, while enterprises mainly consider profit. This chapter uses the Cobb–Douglas production function to calculate household consumer utility and rent to measure enterprise location cost. Consumer utility or location cost is used to calculate the location of households or enterprises. Since different types of households have varying incomes, consumer utility evaluation is conducted by household type and zone. Similarly, as different types of enterprises have different building scale requirements, enterprise location cost evaluation is conducted by enterprise type and zone.

References

Cobb CW, Douglas PH (1928) A theory of production. Am Econ Rev 18(Supplement):139–165
Fan J, Li F (2011) Effect of spatial concentration of manufacturing in China: A review. Southern Econ 6:53–66+6 (in Chinese)
Fujita M (1989) Urban economic theory. Cambridge University Press, Cambridge, UK
Krugman P (1996) The self-organizing economy. Blackwell, Malden, MA
Niu F, Liu W, Feng J (2016) Modeling urban housing price: The perspective of household activity demand. Acta Geogr Sin 71(10):1731–1740 (in Chinese)
Scott AJ (1982) Loctional patterns and dynamics of industrial activity in the modern metropolis. Urban Stud 19:111–142
Shao H (2012) Research on enterprise's location chosen from perspective of linkage cost. East China Econ Manag 5:91–94+108 (in Chinese)
Zheng S, Fu Y, Liu H (2005) An empirical study on the willingness to pay for residential location and location choice of urban residents in China. Econ Geogr 25(2):194–198 (in Chinese)

Chapter 7
Urban Transportation Accessibility Evaluation

This chapter reviews past scholars' understanding and definitions of transportation accessibility and proposes a definition of transportation accessibility. On this basis, a method for evaluating transportation accessibility is established, and finally, a case application of the proposed transportation accessibility evaluation method is conducted.

7.1 Definition of Transportation Accessibility

With the development of the economy and society, the transportation system has gradually improved, and the impact of transportation on economic development has received increasing attention. Transportation accessibility is not only a quantitative measure of the level of transportation development and its advantages but also used to explain spatial changes in social phenomena, such as the spatial structure of land use, urban growth, and the selection of service facilities (Yang and Zhou 1999).

The study of transportation accessibility can be traced back to classical location theory, which characterizes the ease or difficulty of traveling from an origin to a destination (Alonso 1964). As a basic indicator reflecting transportation costs, classical location theory—including agricultural location theory, industrial location theory, and market (urban) location theory—contains theories of accessibility. Christaller coined the German term for "accessibility" in 1933 to discuss central place theory (Christaller 1933). Harris introduced the concept of accessibility by analyzing urban development and transportation planning, considering it the ease of reaching a place, which can be defined through the interaction between different activities and the transportation behavior they induce (Harris and Ullman 1945; Harris 1954). This notion gradually became the foundation of accessibility studies and was applied to urban and transportation planning. Hansen defined accessibility in 1959 as the

F. Niu, *Urban Spatial Evolution Simulation*, Urban Sustainability,
https://doi.org/10.1007/978-981-97-3481-8_7

potential for reaching various opportunities, pioneering the study of the relationship between transportation accessibility and urban land use with the gravity model (Hansen 1959). He believed that the main function of accessibility is to provide opportunities for interaction and exchange in the city. Morris and other scholars viewed accessibility as a measure of the degree of spatial separation of human activities, essentially the ease of reaching activity locations from a given point. They considered accessibility both as an attribute of individuals and spaces, unrelated to actual transportation, measuring the potential to reach selected targets, and as dependent on the existence of opportunities and participation in activities (Morris et al. 1979). Linneker and Spence believed that accessibility refers to the number of opportunities available to people or businesses to reach a place, paying more attention to a region's relative position in the transportation network (Linneker and Spence 1992). Guti regarded accessibility as the possibility of approaching other locations, with improvements in transportation infrastructure leading to higher levels of accessibility measurable through indicators such as weighted average travel time, economic potential, and daily accessibility (Guti 2001). Kwan and other scholars differentiated between place accessibility and individual accessibility, with the former characterizing a location's "approachability" and the latter reflecting the quality of personal life (Kwan et al. 2003).

Domestic scholars' understanding and definitions of accessibility are deeply influenced by international research. Lu Dadao considered accessibility the convenience of exchanging materials, energy, and personnel between a region (country, area, city, linear and point infrastructure) and other related regions, essentially the convenience of socio-economic connections between regions (Lu 1995). The level of accessibility reflects the opportunities and potential for socio-economic and technological exchanges between the region and other related regions, to some extent indicating the region's developmental potential. Continuing Harris and others' definition of transportation accessibility, Yang Jiawen and others summarized the concept of accessibility into three aspects (Harris 1954): the distribution of various geographical entities being evaluated in the region; the types of transportation systems and their combination, service modes, and quality, such as speed, cost, and comfort level; and the socio-economic characteristics of the approach and approached subjects, such as income status, employment situation, and industrial development status.

There are various understandings and definitions of transportation accessibility in the academic community, but they are broadly similar, mainly involving the following aspects: first, transportation costs, including expenses incurred in transportation as well as safety risks and time spent; second, the attractiveness of destinations, reflecting the number of opportunities based on the economic development status and infrastructure conditions of the destination area; third, the choices and characteristics of subjects. This paper defines urban "transportation accessibility" as the convenience of engaging in various economic activities from a certain location. For example, the transportation accessibility of an urban zone i characterizes the convenience of reaching various economic activity opportunities from zone i. Transportation accessibility is evaluated by zoning (spatial units), reflecting the transportation advantage of the zone. Scholars have not reached a consensus on the concept or definition of transportation accessibility (Accessibility).

7.2 Transportation Accessibility Evaluation Methods

Common methods for evaluating transportation accessibility are often limited to a specific number of facilities or a specific range, such as the number of hospitals or companies within a 1-km radius, the distance required to reach 10 hospitals, and the average time or distance to reach a certain number of opportunities. These evaluation methods typically require setting thresholds, but it is challenging to scientifically determine these thresholds in practice, such as whether opportunities within 1 km should be considered effective, and whether opportunities at 1.1 km should be considered. Additionally, when new opportunities are added within the city, the evaluated value of accessibility may decrease, for example, when new employment opportunities are added far away, extending the average time to reach all job positions.

The city is a complete system where various activities interact through transportation, and the transportation conditions of each location are the result of the comprehensive influence of all facilities. Therefore, a reasonable method for evaluating transportation accessibility should consider the distribution of "all city" facilities and the transportation conditions for reaching these facilities. Hence, the core meaning of location conditions should be the quantity and cost of obtaining certain opportunities throughout the entire city, rather than a few facilities or the CBD. In practice, the fundamental reason why transportation facilities (such as subway stations) affect surrounding housing prices is the distribution of economic activities along the transportation line. When the distribution of economic activities along the line changes, it will inevitably affect the housing prices around the transportation facilities. Therefore, if the accessibility of all economic activity opportunities in the city is used to characterize location conditions, it can achieve a dynamic portrayal of location conditions (location conditions will change with the distribution of urban economic activities).

Based on the above discussion, transportation accessibility evaluation involves two aspects: the distribution of various economic activities throughout the city and the transportation conditions for reaching these economic activities. The latter is determined by the supply of transportation facilities and affected by the demand distribution, representing the convenience of transit between zones from a physical transportation perspective. Based on this, this paper establishes a transportation accessibility evaluation model as shown in formula (7.1).

$$A_i = \frac{1}{-\lambda}\left(\ln\left\{\sum_j W_j \exp(-\lambda g_{ij})\right\}\right) \tag{7.1}$$

where A_i characterizes the transportation accessibility of urban zone i; W_j is the number of opportunities within zone j; g_{ij} represents the transportation cost from zone i to j; the distribution coefficient λ indicates the sensitivity of transportation to changes in distance. If transportation is insensitive to distance, λ takes a smaller value; conversely, if transportation is significantly affected by distance, λ should be assigned a larger value. For example, residents traveling by car have lower sensitivity to distance and can tolerate longer distances, so λ can take a smaller value, whereas pedestrians are usually more sensitive to distance and affected more by it, so λ needs to be larger.

In formula (7.1), the coefficient of the distance parameter g_{ij} is negative ($-\lambda$), meaning that as the distance increases, the value of $\exp(-\lambda \cdot g_{ij})$ decreases, diminishing the weight W_j of zone j. Therefore, for zone i, if zone j is difficult to reach (g_{ij} value is high), the opportunities in j become meaningless; conversely, if zone j is easy to reach (g_{ij} value is low), the number of opportunities in zone j has a significant impact. The evaluation result of formula (7.1) makes the dimension of accessibility A_i the same as the transportation cost g_{ij}. Mathematically, the higher the value of A_i, the lower the accessibility.

Formula (7.1) evaluates transportation accessibility using a logarithmic summation method, with the advantage that the model includes both the transportation system and the distribution of economic activities, reasonably characterizing transportation locations and being easy to extend; the formula does not require setting thresholds; if activity opportunities are added far away, the accessibility of zone i does not improve significantly, nor does it decline. Formula (7.1) is an extension and application of McFadden's random utility theory (McFadden 1978). It should be noted that this book uses this model to explain the principles of transportation accessibility evaluation and does not deny the existence of other reasonable mathematical models.

If a computer program is used to calculate the transportation accessibility of all urban zones, the algorithm will include inner and outer loops, where the inner loop is used to calculate the accessibility of a specific block, and the outer loop evaluates the accessibility of all blocks.

The coefficient λ in the formula characterizes the sensitivity of residents' travel to distance. If the city's population is divided into different socio-economic groups, usually, the high-income group has a variety of travel modes, and distance has a relatively small impact on their travel, which can be reflected by assigning a smaller value to λ, reducing the impact of distance on the accessibility evaluation. Correspondingly, formula (7.1) can be modified to formula (7.2).

$$A_i^{seg} = \frac{1}{-\lambda^{seg}}\left(\ln\left\{\sum_j W_j^{seg}\exp\left(-\lambda^{seg} g_{ij}\right)\right\}\right) \tag{7.2}$$

where A_i^{seg} represents the transportation accessibility of zone i for socio-economic group seg, W_j^{seg} is the weight of zone j for socio-economic group seg, and λ^{seg} is the distance coefficient for socio-economic group *seg*, with different values for different socio-economic groups reflecting the varying impact of distance.

For different socio-economic groups, the weight values of zone *j* will also differ. For example, within the same industry, there are differences in the income of job positions, some corresponding to high-income groups and some to middle-low income groups. In other words, the number of opportunities differs for different socio-economic groups, hence the need to determine zone weights by socio-economic group, denoted as W_{j}^{seg}.

Evaluating transportation accessibility in this manner, tailored to socio-economic groups, is theoretically more reasonable and yields more precise results. However, it demands higher data requirements and significantly increases the workload. Its application can be adjusted according to practical needs.

7.3 Multi-Modal Transportation Cost Evaluation

In formula (7.1), the variable g_{ij} represents the transportation cost from zone *i* to zone *j*, effectively forming a matrix that characterizes the urban transportation situation. Transportation costs are typically represented by the minimum of time costs or economic costs. Time and economic costs can be used separately or combined into a generalized cost (g), as shown in formula (7.3a), where $\mathrm{time\_cost_{ij}}$ and $\mathrm{economic\_cost_{ij}}$ are the transportation time cost and economic cost between zones *i* and *j*, respectively.

$$g_{ij} = \ln\left(\exp\left(\mathrm{time\_\ cost}_{ij}\right) + \exp\left(\mathrm{economic\_\ cost}_{ij}\right)\right) \tag{7.3a}$$

Given the presence of multiple transportation modes in cities, such as driving, public transportation (buses, subways), cycling, and walking, and considering that different modes emphasize different aspects of transportation costs, it is necessary to evaluate transportation costs by mode and to aggregate the costs of all modes, as shown in formulae (7.3b) and (7.3c). Here, g_{ijm} is the transportation cost from zone *i* to zone *j* using mode *m*; $time\_cost_{ijm}$ and $economic\_cost_{ijm}$ represent the time and economic costs between zones *i* and *j* for mode *m*, respectively; $\mathrm{W_{time}}$ and $\mathrm{W_{econ}}$ are the weights for time and economic costs, respectively; g_{ij} is the aggregated transportation cost between zones *i* and *j*, incorporating the costs of all modes; λ_{ijm} is the coefficient of transportation cost g_{ijm}, indicating the weight of transportation costs for mode *m*.

$$g_{ijm} = \ln\left(\mathrm{W}_{time} \cdot \exp\left(\mathrm{time\_cost}_{ijm}\right) + \exp\left(\mathrm{W_{econ}} \cdot \mathrm{economic\_cost}_{ijm}\right)\right) \tag{7.3b}$$

$$g_{ij} = \ln \sum_{m} \exp\left(\lambda_{ijm} \cdot g_{ijm}\right) \tag{7.3c}$$

Besides transportation modes, transportation demand and supply also influence the calculation of transportation costs. For example, under the same transportation facility conditions (i.e., the same capacity), high demand along a transportation route can cause congestion and increase transportation costs. Therefore, the evaluation of transportation costs for each mode also needs to consider the situation of transportation demand and supply. Currently, the four-step method is commonly used for urban transportation cost evaluation. Urban transportation cost evaluation can be facilitated by transportation software such as START and TRAM. The result of the transportation cost evaluation is a matrix representing the transportation costs between every pair of zones.

Urban residents can be divided into different socio-economic groups due to income differences. Different socio-economic groups have different priorities for transportation time and distance costs, so the transportation cost gijmgijm should have different values for different socio-economic groups. Correspondingly, the formula for the comprehensive evaluation of urban transportation costs changes to formula (7.4).

$$g_{ij}^{seg} = \ln \sum_{m} \exp\left(\lambda_{ijm}^{seg} \cdot g_{ijm}\right) \tag{7.4}$$

Here, g_{ij}^{seg} represents the comprehensive transportation cost from zone i to zone j for socio-economic group *seg*; λ_{ijm}^{seg} is the weight of the transportation cost for mode m for socio-economic group *seg*.

Moreover, urban transportation trips have various purposes, such as going to school, shopping, or work. Different trip purposes may prioritize transportation costs differently, necessitating further evaluation of transportation costs and accessibility based on trip purposes, using the methods described above.

The aforementioned methods for evaluating transportation accessibility are intended to illustrate the basic principles of transportation accessibility. The specific level of detail to which transportation accessibility evaluation should be refined can be determined based on actual application needs and data availability.

7.4 Activity-Based Transportation Accessibility Evaluation

7.4.1 Transportation Accessibility by Type of Activity

As previously mentioned, this book defines "transportation accessibility" as the convenience of reaching a target (or opportunity) from a certain zone. The targets of transportation trips vary according to the economic activities residents engage in. For instance, in terms of commuting, the transportation accessibility of block i refers to the ease of reaching employment opportunities from zone i, determined by the distribution of all job positions in the city and the transportation conditions between them and zone i; similarly, this applies to activities such as education, healthcare, and consumption. Since the spatial distribution patterns of various economic activity opportunities differ, the results of transportation accessibility evaluations will inevitably vary. In other words, the same location has different values for different economic activities. The transportation accessibility of a zone reflects the influence of the city's economic activities on that block.

Therefore, urban transportation accessibility should be evaluated by the type of activity. Assuming that city economic activities are divided into employment, education, consumption, and healthcare, evaluating the accessibility of each type of activity using formula (7.1) requires setting the weights (W_j) for each zone. To evaluate employment accessibility, the number of job positions in each block can be used as weights, as shown in formula (7.5). Here, A_i^j represents the employment accessibility of zone i, indicating the convenience of commuting from zone i, and W_j is the weight of zone j, namely the number of job positions distributed within zone j.

$$A_i^j = \frac{1}{-\lambda^J}\left(\ln\left\{\sum_j W_j \exp(-\lambda^j g_{ij})\right\}\right) \tag{7.5}$$

7.4.2 Origin Accessibility (Active Accessibility) and Destination Accessibility (Passive Accessibility)

In the context of transportation, there are origins and destinations. As "origins" (typically homes), the focus is on the convenience of reaching various opportunities, termed "origin accessibility." Conversely, as "destinations" (typically businesses), the concern is about the convenience of being reached, termed "destination accessibility." Since "origins" (homes) initiate transportation, origin accessibility is also referred to as active accessibility. Correspondingly, destination accessibility is referred to as passive accessibility. The definitions of opportunities differ between "origin accessibility" and "destination accessibility." For example, the convenience for residents to go shopping represents origin accessibility, while the convenience of reaching a shopping center represents destination accessibility. The former focuses on the distribution of shopping centers, while the latter focuses on the distribution of the population (consumers). The former influences home location choices, and the latter influences the siting of malls or supermarkets.

Accordingly, the previously discussed transportation accessibility evaluation falls under origin accessibility evaluation. Taking shopping (consumption) activities as an example, the method for evaluating destination accessibility considers the spatial distribution of residents (consumers) and transportation conditions, as shown in formula (7.6). Here, A_j^C represents the destination accessibility of zone j for shopping (consumption) activities, and W_i is the weight of zone i, representing the number of consumers in the zone.

$$A_j^C = \frac{1}{-\lambda^c}\left(\ln\left\{\sum_i W_i \exp(-\lambda^c g_{ij})\right\}\right) \tag{7.6}$$

Similarly, weights W_i can be replaced with employment, school-age population, or population numbers (patient numbers) to evaluate the destination accessibility for employment, education, healthcare, and other economic activities. As a result, each zone has four types of accessibility, corresponding to eight evaluation values, representing the origin and destination accessibility for the four types of activities. The classification of transportation accessibility changes with the classification of economic activities.

7.5 Case Study Application

This section explains the application of origin and destination accessibility in urban spatial analysis.

7.5.1 Study Area and Data

Beijing, as a megacity with a developed economy, large population, and complex transportation, is a typical case study area for urban research. The main urban area of Beijing is located within the Sixth Ring Road, with four other outlying districts (Huairou, Miyun, Pinggu, Yanqing) outside the Sixth Ring Road. To comprehensively explore urban spatial evolution and expansion, the entire city of Beijing is taken as the case study area, focusing on the 12 districts within (and intersecting with) the Sixth Ring Road at the street (township) scale. The four outlying districts, which are mountainous, are not further subdivided spatially but are considered at the county level. The study area thus comprises 243 zone, as shown in Fig. 7.1. It should be noted that this study focuses on economic elements, and the use of county-scale data for some suburban areas does not affect simulation results.

The study utilized four sets of data: housing price data; road network data digitalized based on the 2010 map (including highways, urban expressways, national roads, provincial roads, county roads, subway lines); the Sixth Population Census data (household population data); and Beijing's economic activity distribution data. Housing price data were obtained from real estate brokerage companies (SouFun website), averaging the housing prices within each zone to represent the zone's housing price. The distribution data of Beijing's economic activities were acquired by first obtaining the scale and other relevant information of various enterprises based

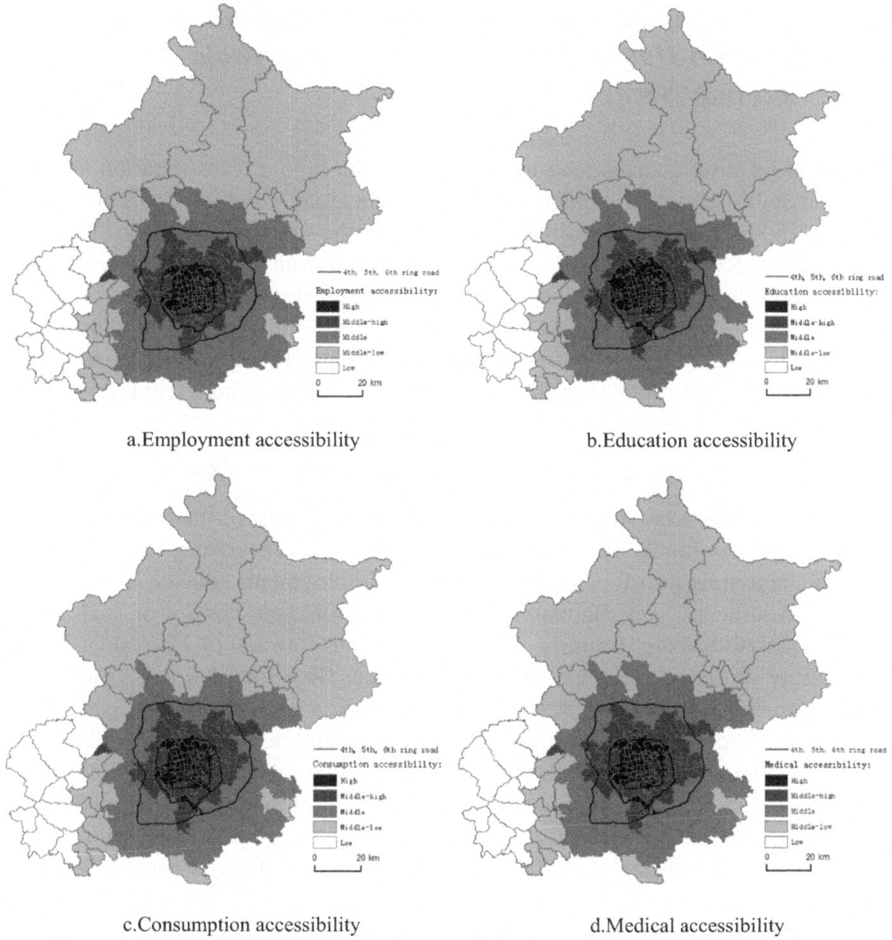

Fig. 7.1 Activity-based transportation accessibility

on economic census data and field research. Then, using the map remote access interface (API) and POIs (Points of Interest) provided by electronic map providers, the spatial coordinates of each enterprise were determined. Over 700,000 enterprise records were obtained, covering nearly all companies, schools, research institutes, hospitals, and other workplaces in Beijing, including spatial location, asset scale, number of employees, etc. This resulted in detailed Beijing employment distribution data to represent the distribution of economic activities.

7.5.2 Application of Origin Accessibility—Simulation of Urban Housing Prices Based on Household Location Needs

Household location choices tend to favor areas with higher transportation accessibility, inevitably leading to higher housing prices in areas with high accessibility. As discussed, household accessibility is determined by the spatial distribution of various enterprises. Since household members engage in different economic activities, the choice of household location is determined by the accessibility of the economic activities engaged in by the household members. This case study will use activity-based transportation accessibility, combined with the composition of household members, to comprehensively evaluate household accessibility, representing various household location needs, and thereby simulate urban housing prices.

1. **Activity-Based Urban Transportation Accessibility Evaluation**

The transportation accessibility of a zone reflects the impact of the city's economic activities on that block, indicating the block's value. This study divides urban economic activities into four major categories: employment, education, consumption, and healthcare, and calculates the transportation accessibility of each block for these activities using formula (7.1). According to formula (7.1), evaluating the accessibility of zone i requires setting the weights (W_j) for each zone j to represent the number of economic activity opportunities in each zone. This paper uses the scale of employment in various industries as weights, which well reflects the number of opportunities for each type of economic activity. Therefore, the evaluation of employment accessibility uses the total number of job positions in each zone, while the evaluations of education, healthcare, and consumption accessibility use the number of employees in education, healthcare, and other services as weights, respectively.

Evaluating transportation accessibility with formula (7.1) also requires solving for urban transportation costs (g_{ij}). This study overlays urban highways, expressways, national roads, provincial roads, county roads, and subway lines to form a network, assigning different speed values to each level of transportation route to solve for the shortest transportation time between zones (this study does not evaluate transportation by mode). After determining weights (W_j) and transportation costs (g_{ij}), formula (7.1) is used to evaluate various types of transportation accessibility, as shown in Fig. 7.1. The accessibility of each category reflects the locational advantages of the zone.

As shown in Fig. 7.1, we divide each type of accessibility into five levels: high, medium–high, medium, medium–low, and low. The spatial pattern of various accessibilities is generally similar, that is, higher in the central city area and decreasing towards the suburbs. This is because the density of economic activities in Beijing currently decreases from the central city area to the suburbs. The evaluation results indicate: (1) For employment trips, the area within the Fourth Ring Road still has the highest locational advantage, indicating that the majority of employment opportunities are located in the main urban area within the Fourth Ring Road, which also has convenient transportation facilities; (2) For educational trips, the advantage is significant within the Fifth Ring Road, especially in the northwest, indicating that Xicheng and Haidian districts have excellent educational resources, making educational trips extremely convenient in these areas; (3) The distribution of consumption accessibility and healthcare services accessibility is generally similar, concentrated especially within the Fourth Ring Road, stemming from past development patterns.

2. **Evaluation of Household Location Needs**

Urban residents typically choose living locations by household unit, and a household's location choice reflects the combined location needs of its members. Household members can generally be divided into children, working-age population, and elderly (retirees). Each member type may further be subdivided (e.g., working-age population into employed and unemployed), which is not discussed in this book. The activities engaged in by different household members vary, determining their different needs, for example: children require education, working-age individuals require employment, and the elderly need healthcare services, in addition to the household's need for convenient consumption trips. Due to the different compositions of household members, their transportation needs vary. For instance, the location choice of a couple mainly considers employment accessibility, while a family with children will consider both employment and education accessibility; families with elderly members will also consider accessibility to healthcare services. Thus, household location choice should be a comprehensive consideration of the travel needs of all members, leading to the concept of household accessibility, which represents the transportation convenience of a zone for a particular type of household. Household accessibility evaluation requires considering the structure of household members and weighting various types of accessibility, such as employment, education, consumption, and healthcare.

Based on the definition of household accessibility, the accessibility of a zone represents its value to households. The evaluation of household accessibility requires determining the weights of various economic activities for the household based on the composition of household members and weighting the various accessibilities accordingly. This means that the same block has different values for different households. The Sixth Beijing Population Census classifies households into single-generation, two-generation, three-generation, four-generation or more households, and provides information on children, working-age population, and retired population. Based on this, the average member structure of various types of households can be calculated, as shown in Table 7.1.

This study calculates the weights of various types of transportation accessibility for different types of households based on the travel frequency of household members, as shown in formula (7.7). Here, *Weight* represents the weight of a certain type of economic activity, i.e., the weight of that type of accessibility. *Person* is the number of a certain type of members, and *Frequency* is the frequency of these members engaging in a certain activity (number of times), with the constant

Table 7.1 Demographic structure of households

Household	Household proportion (%)	Children (per household)	Workers (per household)	Retired (per household)
1-generation household	50.4	0.34	1.34	0.16
2-generation household	39.4	0.71	2.71	0.25
3-generation household	10.1	1.09	4.11	0.32
4-and-plus generation household	0.2	1.47	5.56	0.32

2 accounting for round trips.

$$Weight = Person * Frequency * 2 \tag{7.7}$$

For employment and educational activities, *Person* corresponds to the number of working family members or minors, respectively, with an average of 23.7 working days per month calculated based on 5 working days per week, resulting in a monthly activity frequency (*Frequency*) of 23.7 times per person; for consumption activities, it is calculated as 4 times per month (once a week) for each family; for healthcare activities, *Person* refers to the number of retired members, with *Frequency* being the rate of medical visits per person per month, calculated as 2 times. Based on this, the weights of various activities for different types of families are as specified in Table 7.2.

Using these weights, the household accessibility is obtained by weighting the accessibility of various economic activities, as shown in formula (7.8). Here, A_i^v represents the household accessibility of zone i for family type v; $A(J)_i$, $A(E)_i$, $A(S)_i$, and $A(M)_i$ are the employment, education, consumption, and medical accessibility

Table 7.2 Weights of various kinds of economic activities on housing location

Household	Weight of employment	Weight of education	Weight of consumption	Weight of medical services
1-generation household	58.15	14.82	3.68	0.32
2-generation household	117.66	30.99	7.35	0.50
3-generation household	178.50	47.10	11.03	0.64
4-and-plus generation household	241.26	63.87	14.71	0.65

of zone *i*, respectively; $w(J)_v$, $w(E)_v$, $w(S)_v$, and $w(M)_v$ are the weights of each type of accessibility for family type *v*.

$$A_i^v = A(J)_i \cdot w(J)_v + A(E)_i \cdot w(E)_v + A(S)_i \cdot w(S)_v + A(M)_i \cdot w(M)_v \quad (7.8)$$

Each zone has four household accessibility evaluation results, representing the transportation accessibility of the zone for single-generation, two-generation, three-generation, and four-or-more-generation households. A zone suitable for single-generation household location might not be suitable for two- or three-generation households.

From Fig. 7.2, the distribution of household accessibility generally decreases from the central city area to the suburbs. This is due to the concentration of various economic activities in the central city area, with density gradually decreasing towards the suburbs. However, the household accessibility evaluation value for each zone is not the same. For housing development, housing types can be determined based on family needs at different locations, e.g., locations with strong demand from single-generation households are suitable for developing small housing units, while locations with high accessibility for four-or-more-generation households are suitable for developing large housing units. This study provides a reliable reference. The focus of this paper is on simulating housing prices, and it does not delve into the spatial differences in various types of accessibility.

3. **Simulation of Housing Prices Based on Household Location Needs**

Household accessibility reflects the demand level of families for zones, influencing housing prices. After preliminary analysis, household accessibility and housing prices distribution exhibit a power function relationship, hence, we set the function form as shown in formula (7.9).

$$P_i = HACC_{i,1}^{m1} \cdot HACC_{i,2}^{m2} \cdot HACC_{i,3}^{m3} \cdot HACC_{i,4}^{m4} * \varepsilon \quad (7.9)$$

where P_i is the housing price of zone *i*, $HACC_{i,j}$ is the household accessibility of zone *i* for family type *j*, and *mi* is a to-be-determined exponent. The function is first logarithmized to convert it into a linear function, and then regression analysis is performed using the OLS method, with results as shown in Table 7.3. The results show that the model achieved an ideal fit within the whole city, indicating that household accessibility can well reflect family needs. This study provides a good approach for the simulation and analysis of housing prices.

According to Table 7.3, the household accessibility of single-generation, two-generation, and three-generation households is not significant in the variables. However, the household accessibility for households with four or more generations shows a very high correlation with the distribution of housing prices. When the number of family members is small, their travel needs are relatively singular. In contrast, households with four or more generations have more members and comprehensive travel needs, requiring consideration of various demands in their

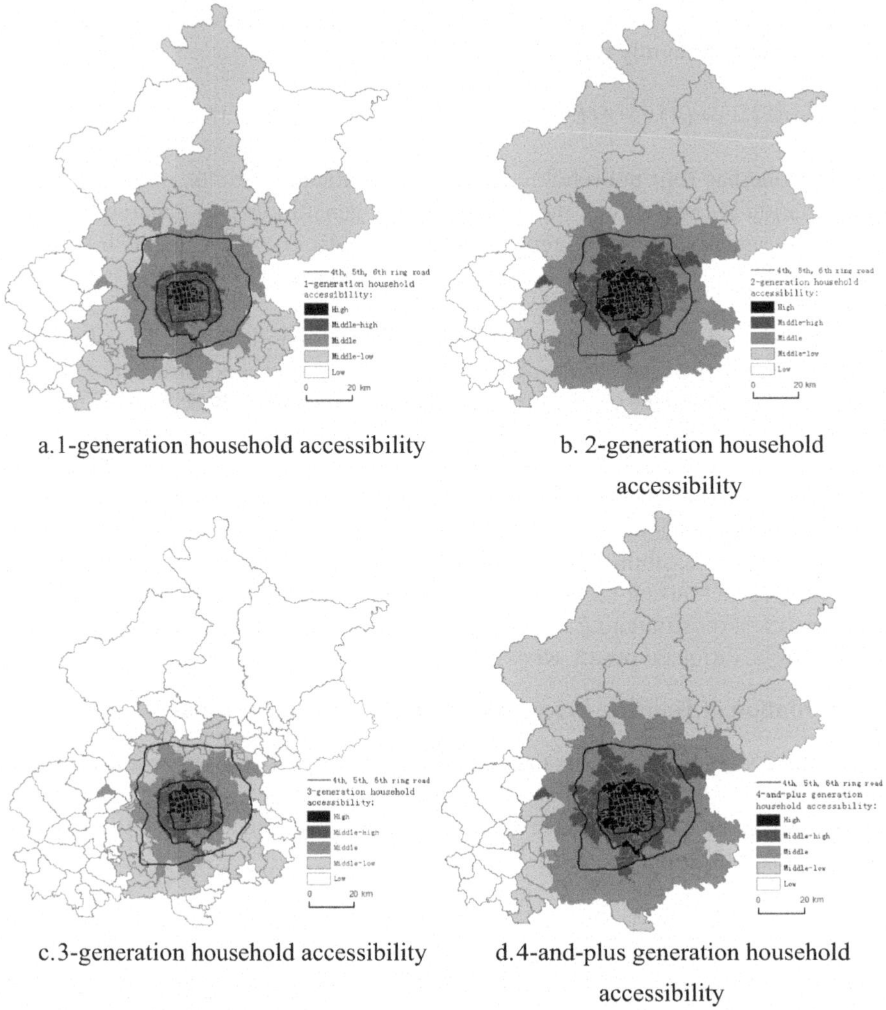

a. 1-generation household accessibility

b. 2-generation household accessibility

c. 3-generation household accessibility

d. 4-and-plus generation household accessibility

Fig. 7.2 Household accessibility

location choices. These results confirm that urban housing prices are the outcome of the combined effects of various economic activities. Although two-generation households are larger than single-generation households, the accessibility of two-generation households was excluded in the regression. Two-generation households often have children attending school (education activity), and their residential location choices are greatly influenced by the distribution of educational resources. This result indicates that, across the city, the distribution of educational resources is not a key factor affecting housing prices. This could be because only a few key schools can raise nearby housing prices, but their impact is limited to a small area around the

Table 7.3 The estimation results of household accessibility and housing price

Variables	Descriptions	Exponentials	Values of Exponentials	T-test
$Val_{i,1}$	Household accessibility of 1-generation household in zone *i*	m1	0.002	Insignificant
$Val_{i,2}$	Household accessibility of 2-generation household in zone *i*	m2	–	Insignificant
$Val_{i,3}$	Household accessibility of 3-generation household in zone *i*	m3	0.004	Insignificant
$Val_{i,4}$	Household accessibility of 4-and-plus generation household in zone *i*	m4	0.74	17.20**

Note $R^2 = 0.70$, F = 296, "**" means significant at 1% level

school, i.e., within the school district. Furthermore, under the previous family planning policy, most families had only one child, making school education a temporary need for families.

4. **Conclusion and Discussion on Transportation Accessibility and Housing Prices**

Cities are organic wholes where transportation and land use interact. Existing studies often use location relationships to depict the spatial heterogeneity of housing prices, paying less attention to the relationship between housing prices and the city's transportation system and the distribution of economic activities. The above case study, based on transportation accessibility, discussed the intrinsic connection between urban transportation, economic activities, and housing prices, demonstrating the effectiveness of the activity-based transportation accessibility evaluation method.

First, the transportation accessibility evaluation model constructed in this paper comprehensively considers the distribution of citywide economic activities and transportation conditions, evaluating transportation accessibility by "activity." The model reflects that the accessibility of each zone is the result of the combined action of all economic activities in the city, embodying the basic laws of urban activities interaction.

Second, a method for evaluating household accessibility was established, assessing household location needs based on family structure. Families are the basic units of housing choice, and starting from the perspective of family needs facilitates a systematic and comprehensive study of the factors affecting urban housing. The location needs of a family are a composite reflection of the location needs of all its members, making the value of the same zone different for different families. This model can provide developers with decision-making references on what types of housing to develop in different zones.

Next, the study simulated urban housing prices based on family needs, achieving good model effects across the city. The locational needs of households with four or more generations are significantly correlated with the spatial distribution of housing prices. Households with four or more generations are large families with a variety of travel needs, which also confirms that housing prices are influenced by various economic activities. The model provides a good approach for simulating urban housing prices.

This paper classified economic activities into four major categories: employment, education, consumption, and healthcare, without further subdivision due to data limitations. Future research could further subdivide these categories based on application needs. When calculating transportation costs, the study did not consider the impact of transportation supply and demand, such as road capacity and traveling population. Moreover, the study considered only a single mode of transportation, while in reality, different modes (e.g., public transit, driving) have different sensitivities to distance. Thus, constructing a more sophisticated transportation model to calculate the comprehensive transportation costs of the city is a direction for further research.

7.5.3 Destination Accessibility

Unlike household accessibility, corporate location choices focus on destination accessibility. We evaluated the destination accessibility of employment, education, healthcare, and consumption activities for businesses as classified above. When evaluating the destination accessibility of zone j using formula (7.6), it's necessary to set weights (W_i) for all zones, representing the number of opportunities in each block. Employment accessibility evaluation used the number of working populations in each block, while education, healthcare, and consumption accessibility evaluations used the number of school-age populations, elderly populations, and total population (number of consumers), respectively. Figure 7.3 shows the evaluation results of destination accessibility for employment, education, healthcare, and consumption activities in Beijing, representing the locational advantages of various types of businesses. Corresponding to Fig. 7.3, Fig. 7.1 previously presented origin accessibility, i.e., active accessibility. Destination accessibility affects commercial property prices, which is not further analyzed through regression in this paper.

7.6 Chapter Summary

This chapter defines "transportation accessibility" in the city as the convenience of engaging in various economic activities from a certain location and establishes an activity-based method for evaluating transportation accessibility. This method considers that the transportation accessibility of any location in the city is determined collectively by all opportunities citywide. The activity-based transportation accessibility evaluation method assesses transportation accessibility based on types of city activities, reflecting the ease of reaching various activity opportunities and

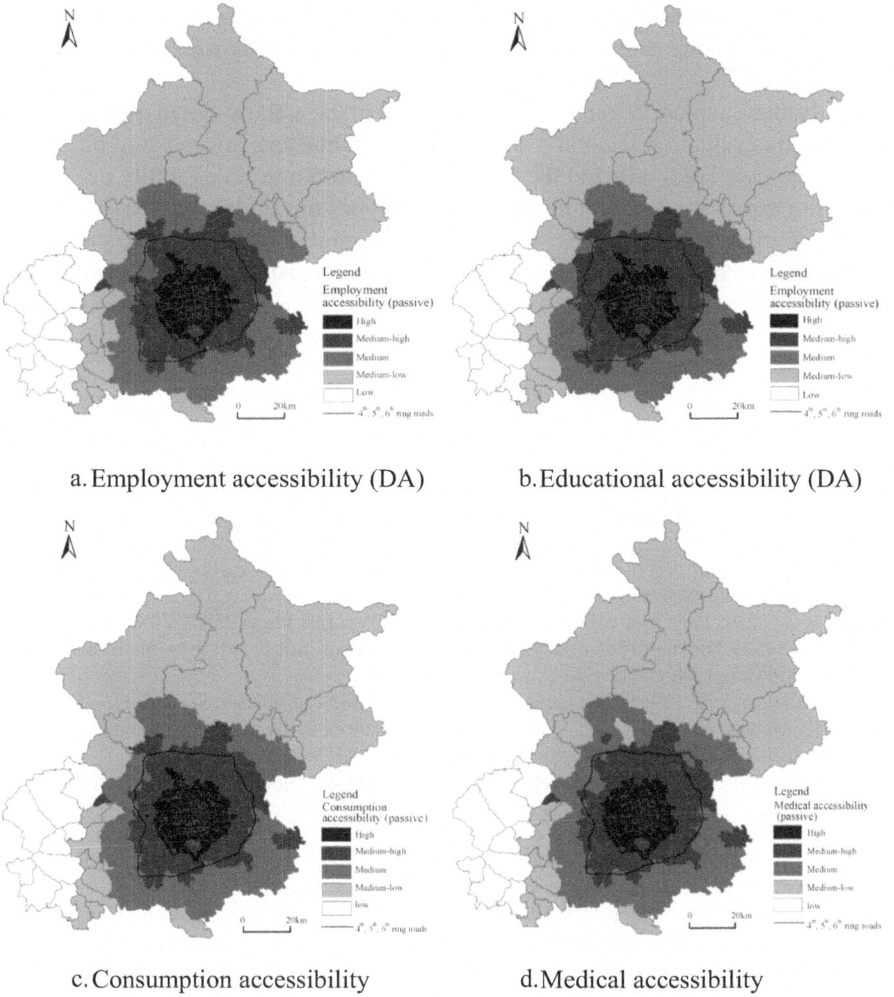

Fig. 7.3 Destination accessibility in various categories in Beijing City

comprehensively evaluating the accessibility for different types of families based on the structure of household members.

This paper introduces the concepts of origin accessibility and destination accessibility. The former represents the convenience of reaching various economic activity opportunities from the origin, while the latter indicates the convenience of being reached as a destination within a zone. Origin accessibility has different definitions for opportunities based on the type of activity, such as educational accessibility (active) focusing on the distribution of educational opportunities. Similarly, destination accessibility has its own definitions for opportunities, for example, supermarkets or malls focus on the distribution of the population (consumers). Origin accessibility affects housing prices, whereas destination accessibility impacts commercial property prices.

Lastly, this chapter presents a case application of the activity-based transportation accessibility evaluation method—simulating urban housing prices based on family needs. The case study utilized the proposed transportation accessibility evaluation method to assess the location needs of various types of families and based on this, simulated urban housing prices, achieving good simulation results. The simulation outcomes indicate that a key factor affecting urban housing prices is employment accessibility.

References

Alonso W (1964) Location and land use. Harvard University Press, Cambridge, MA

Christaller W (1933) Central place in Southern Germany. Baskin C (trans) (1966). Prentice Hall, London, 1966

Guti J (2001) Location, economic potential and daily accessibility: an analysis of the accessibility impact of the high-speed line Madrid ± Barcelona ± French border q. J Transport Geogr 14

Hansen WG (1959) How accessibility shapes land use. J Am Inst Plann 25:73–76

Harris CD (1954) The market as a factor in the localization of industry in the United States. Ann Assoc Am Geogr 44(4):315–348

Harris CD, Ullman EL (1945) The nature of cities. Ann Am Acad Pol Soc Sci 242(1):7–17

Kwan MP, Murray AT, Tiefelsdorf M (2003) Recent advances in accessibility research: representation, methodology and applications. J Geogr Syst 5(1):129–138

Linneker BJ, Spence NA (1992) An Accessibility analysis of the impact of the M25 London orbital motorway on Britain. Reg Stud 26(1):31–47

Lu D (1995) Regional development and its spatial structure. Science Press, Beijing (in Chinese)

McFadden D (1978) Modelling the choice of residential location. In: Karlquist A et al (eds) Spatial interaction theory and residential location. North Holland, Amsterdam, pp 75–96

Morris JM, Dumble PL, Wigan MR (1979) Accessibility indicators for transport planning. Transp Res Part a: General 13(2):91–109

Yang J, Zhou Y (1999) Accessibility: concept, measurement, and application. Geogr Nat Territory Res 02:62–67 (in Chinese)

Chapter 8
Urban Activity Location Simulation

Following the groundwork laid in previous chapters, this chapter will further elaborate on the construction process of the ActSim model. The location model discussed in this chapter is a core part of the ActSim model. It integrates the sub-models described in earlier chapters to determine the spatial distribution patterns of urban households and businesses and establishes a computer algorithm to simulate the urban spatial evolution process.

8.1 Urban Activity Location Model

The location model first comprehensively evaluates the location conditions of each zone based on location costs and transportation accessibility (choosing these two most important factors as examples). It then determines the locations of households and businesses according to location conditions, combined with the distribution of housing.

8.1.1 Location Evaluation

The ActSim model, based on "activity," evaluates locations for various "activities" separately. The location choice of households is the result of a comprehensive evaluation of various influencing factors (Niu et al. 2016). The household location cost and household accessibility are integrated into an index representing the value of a location, termed household location utility. Household location utility reflects the value or attractiveness of a zone to households. If households are subdivided into multiple types, the evaluation of household location utility is conducted separately for each type of household, as shown in Formula (8.1a).

F. Niu, *Urban Spatial Evolution Simulation*, Urban Sustainability,
https://doi.org/10.1007/978-981-97-3481-8_8

$$V_{t+1,i}^{H} = \theta^{U} \cdot U_{t+1,i}^{H} + \theta^{A} \cdot A_{t+1,i}^{H} \tag{8.1a}$$

where $V_{t+1,i}^{H}$ is the location utility of zone i for type H households in year $t+1$, $U_{t+1,i}^{H}$ is the consumption utility of zone i for type H households in year $t+1$, $A_{t+1,i}^{H}$ is the transportation accessibility (household accessibility) of zone i for type H households in year $t+1$, θ^{U} and θ^{A} are coefficients representing the weights of each index.

Similarly, the indices for businesses are integrated to form an index representing the value of a location, termed business location utility. Business location utility reflects the value or attractiveness of a zone to businesses. If businesses are subdivided into different types, the business location utility needs to be evaluated separately for each type of business, as shown in Formula (8.1b).

$$V_{t+1,i}^{e} = \theta^{C} \cdot C_{t+1,i}^{e} + \theta^{A} \cdot A_{t+1,i}^{e} \tag{8.1b}$$

where $V_{t+1,i}^{e}$ is the location utility of zone i for type e businesses in year $t+1$, $C_{t+1,i}^{e}$ is the location cost of zone i for type e businesses in year $t+1$, $A_{t+1,i}^{e}$ is the accessibility (destination accessibility) of zone i for type e businesses in year $t+1$, θ^{C} and θ^{A} are coefficients representing the weights of each index.

8.1.2 Urban Activity Location Model

The urban activity location model is used to determine the spatial distribution pattern of urban activities, i.e., the distribution of households and businesses.

1. **Household Location Model**

The household location model calculates the spatial distribution of urban households, determining the number of households in each zone. The model determines household distribution based on urban household location utility (V). Since household site selection considers not only the absolute values of influencing factors but also their trends, we model based on the change (increment) in location utility.

Furthermore, according to the agglomeration effect, the location choices of various types of households are influenced by the existing household distribution and the distribution of housing. Since researchers cannot know how each individual evaluates each location, we chose the discrete choice model (McFadden 1978). The discrete choice model assumes that the error terms are independent and uniformly distributed. The household location model is shown in Formula (8.2).

$$H(L)_{t+1,i}^{H} = H(M)_{t+1}^{H} \cdot \frac{H_{ti} \cdot F(A)_{t+1,i} \cdot \exp\left(\Delta V_{t+1,i}^{H}\right)}{\sum_{i} \left\{H_{ti} \cdot F(A)_{t+1,i} \cdot \exp\left(\Delta V_{t+1,i}^{H}\right)\right\}} \tag{8.2}$$

where $H(L)_{t+1,i}^{H}$ is the number of type H households migrating into zone i in period $t+1$; $\Delta V_{t+1,i}^{H}$ is the change in location utility of zone i for type H households in period $H(M)_{t+1}^{H}$ is the total number of type H households relocating within the city in period $t + 1$; H_{ti}^{H} is the number of type H households in zone i in period t; $F(A)_{t+1,i}$ is the number of available housing units (area) in period $t + 1$, without further classification of housing in this book.

2. **Business Location Model**

The business location model determines the distribution of businesses, i.e., the distribution of economic activities. This book represents the scale of economic activities by the number of employees rather than the number of businesses. The location choices of businesses are determined by business location costs, transportation accessibility (destination accessibility), and are also influenced by the existing business distribution and commercial property supply. Based on this, the business location model is established using the discrete choice model, as shown in Formula (8.3).

$$E(L)_{t+1,i}^{e} = E(M)_{t+1}^{e} \cdot \frac{E_{ti}^{e} \cdot F(A)_{t+1,i} \cdot \exp\left(\Delta V_{t+1,i}^{e}\right)}{\sum_{i} \left\{E_{ti}^{e} \cdot F(A)_{t+1,i} \cdot \exp\left(\Delta V_{t+1,i}^{e}\right)\right\}} \tag{8.3}$$

where $E(L)_{t+1,i}^{e}$ is the scale (number of employees) of type e businesses migrating into zone i in period $t + 1$; $\Delta V_{t+1,i}^{e}$ is the change in location utility of zone i for type e businesses in period $t + 1$; $E(M)_{t+1}^{e}$ is the scale (number of employees) of type e businesses needing to relocate within the city in period $t + 1$; E_{ti}^{e} is the number of type e businesses in zone i in period t; $F(A)_{t+1,i}$ is the number of available business premises (area) in zone i in period $t + 1$, similar to housing, the book does not further classify business premises (commercial properties).

8.2 Simulation of Urban Spatial Evolution Process

8.2.1 Urban Spatial Evolution Simulation Algorithm

The location model outputs the distribution patterns of households and businesses, i.e., the land use patterns. Based on the theory of interaction between urban land use and transportation, the urban spatial evolution process is an interaction between land use and transportation systems, where the transportation system affects the land use system (urban activity distribution) through accessibility, and vice versa, both influencing each other and moving towards an equilibrium state. Any change in factors will lead to the urban system moving towards a new equilibrium state. To simulate the above process, a computer algorithm has been developed, as shown in Fig. 8.1.

The algorithm depicted in Fig. 8.1 is an integrated application of the aforementioned models. The algorithm concept is as follows:

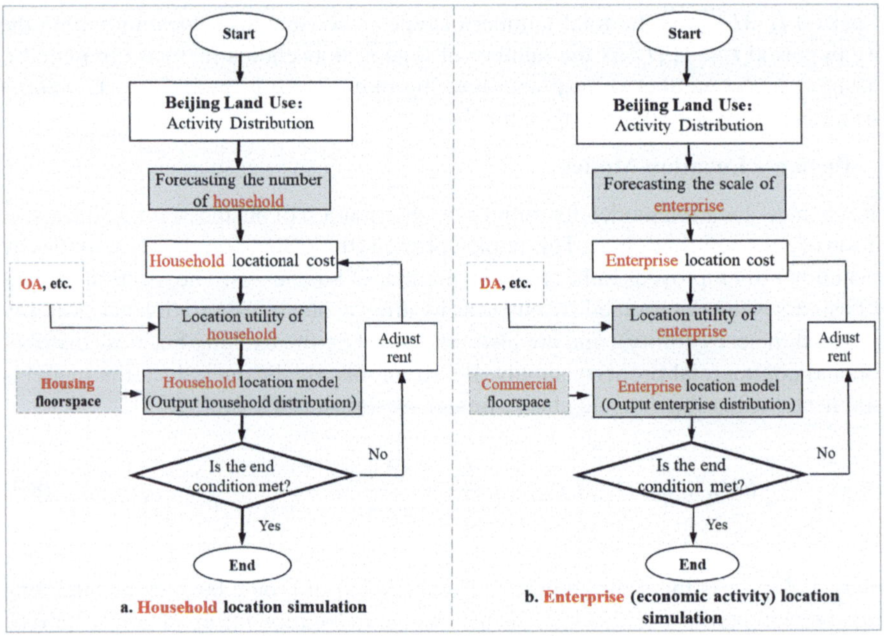

Fig. 8.1 ActSim model algorithm

1. The transportation model calculates urban transportation accessibility for the year t based on the distribution of socio-economic activities and transportation costs, including origin and destination accessibility for various activities.
2. It simulates the transition process of urban households and businesses, predicting the scale of households and businesses.
3. It calculates the consumption utility for each zone based on rent and household income and determines the location costs for businesses based on rent.
4. It calculates the location utility of each zone based on transportation accessibility and consumption utility/location costs.
5. Using the location utility and housing distribution, the location model calculates the spatial distribution of households/businesses, changing the urban activity distribution density.
6. Changes in the distribution of households/businesses lead to changes in rent, adjusting the rent and repeating the above process until the termination condition is met. The urban spatial state at this point is the predicted value for the year $t + 1$.

The termination condition in the algorithm refers to when the results of two consecutive cycles show no change or very little change. Using the predicted values for the year $t + 1$ and setting policy scenarios can further predict the spatial distribution patterns of households and businesses for the year $t + 2$ and so on, predicting the future urban spatial conditions year by year. In this algorithm, changes in either land use (housing patterns) or transportation facilities will lead to the urban system moving towards a new equilibrium state.

The algorithm implicitly includes a rent adjustment model. Rent (housing prices) is one of the key factors affecting urban location utility. After the location model calculates the distribution pattern of urban households or businesses, the spatial distribution of housing demand changes, and rent inevitably changes. The rent model is based on the demand and supply of housing (Albouy and Ehrlich 2014; Mumtaz 1995), taking into account the rent from the previous year to simulate new rent. The new rent is then used to calculate the spatial distribution pattern of households or businesses. The greater the demand for housing in a zone (the more households or businesses), the higher the rent. Formula (8.4a) is the housing rent model.

$$r'^{h}_{pi} = r^{h}_{pi}\left[\frac{a^{h}_{pi} \cdot H(L)_{pi}}{F(A)^{h}_{pi}}\right] \tag{8.4a}$$

where $r\prime^{h}_{pi}$ is the new housing rent within zone i; r^{h}_{pi} is the rent calculated in the previous cycle (or the rent for the year t if it is the first cycle); a^{h}_{pi} is the distribution density of households within zone i (actually the reciprocal of density), i.e., the average housing area per household; $H(L)_{pi}$ is the number of households moving into zone i; $F(A)^{h}_{pi}$ is the total area of available housing within zone i.

By replacing variables related to households with those related to businesses in Formula (8.4a), we get the business premises rent model, as shown in Formula (8.4b). Where r'^{e}_{pi} is the new business premises rent within zone i; r^{e}_{pi} is the rent calculated in the previous cycle (or the rent for the year t if it is the first cycle); variables a^{e}_{pi} represent the distribution density of businesses within zone i (actually the reciprocal of density), i.e., the average office area per employee; $E(L)_{pi}$ is the number of businesses (number of employees) moving into zone i; $F(A)^{e}_{pi}$ is the total area of available business premises within zone i.

$$r'^{e}_{pi} = r^{e}_{pi}\left[\frac{a^{e}_{pi} \cdot E(L)_{pi}}{F(A)^{e}_{pi}}\right] \tag{8.4b}$$

8.2.2 Termination Condition and Convergence of the Algorithm

The termination condition refers to no change or very minor changes between two cycles of the algorithm. A threshold must be pre-set, which can be determined based on the specific needs of the application.

The algorithm ultimately converges to an equilibrium state, requiring accurate and reasonable input data to ensure the algorithm's convergence. This issue, though technical in appearance, is fundamentally determined by the reasonableness of the mathematical model and data. On one hand, it's essential to ensure that each sub-model is scientifically valid; on the other hand, input data must conform to economic laws. For example, if the city has abnormally high rents that don't conform to economic principles, inputting such data into the system could cause the algorithm to oscillate and fall into an infinite loop. To avoid this, it's necessary to use scientific pre-processing methods to correct unreasonable data while ensuring the authenticity of the data is not lost.

8.2.3 Frequency of Interaction Between Land Use and Transportation Models

Taking household distribution simulation as an example, the algorithm depicted in Fig. 8.1 uses origin accessibility when calculating household distribution. According to the transportation accessibility evaluation method, origin accessibility is determined by business distribution, indicating that business distribution affects household location choices; similarly, the business location simulation uses destination accessibility, which is determined by household distribution, thus the model also reflects how urban household distribution affects business site selection. This illustrates the spatial interaction within the city that the algorithm captures.

The algorithm's output predicts values for the next year. During the cycle, the same accessibility evaluation value is read, meaning the transportation model to calculate transportation accessibility is not repeatedly called. After determining household distribution, the algorithm calculates business distribution without alternating operations. In terms of the frequency of interaction between transportation and land use models, recalculating transportation accessibility after each cycle would make the model interaction too frequent, affecting the running time of the algorithm. It is believed that from the perspective of long-term prediction results, there would not be significant differences, thus frequent model interaction is deemed unnecessary.

8.2.4 Spatial and Temporal Scales of the Model

ActSim is used to simulate the urban spatial development process, predicting the distribution pattern of socio-economic activities. To describe urban space, it's necessary to divide the urban area into different spatial units, considering each unit internally homogeneous. The initial case city for the LUTI model was Pittsburgh (Lowry 1964). This city, like subsequent applications in developed countries, is relatively small. In contrast, large Chinese cities, such as Beijing and Shanghai, usually cover a larger area, and the division of urban spatial units directly affects the model's operational efficiency and the difficulty of data collection. If spatial units are too small, although the prediction results are more accurate, it would lead to a huge computational burden and difficulty in data acquisition. Statistical data for Chinese cities are mostly at the administrative division level, and data below the county level are hard to obtain, which also hinders the application of LUTI models in China.

In terms of the temporal scale, since statistical data are usually annual, it's common to use 1 year as the minimum time granularity. In practice, the time granularity can be adjusted based on application needs, such as every 5 years as a time unit. Regardless of the chosen time granularity, model calibration should be performed with data of the same granularity. For example, if the model predicts the future urban condition on an annual basis, historical annual data should be used for model calibration.

The subsequent case applications of the ActSim model in this book focus on the internal space of Beijing city. Considering data accessibility and operational feasibility, the study units are streets (towns, villages), i.e., zones. The collection and preparation of socio-economic activity spatial data are conducted at the street (town, village) level. The time sequence adopts natural years as the time granularity, predicting annually, as depicted in Fig. 8.2.

In the land use and transportation interaction process shown in Fig. 8.2, the transportation model evaluates city transportation accessibility based on land use and the transportation network for year *t*, then the location model uses the land use data and transportation accessibility results for year *t* to predict the distribution of households and businesses for year $t + 1$. The location model, during the recursive operation, does not recalculate transportation accessibility with the transportation model.

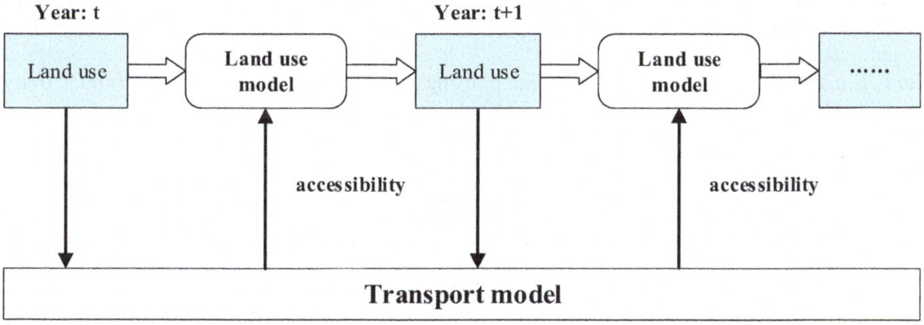

Fig. 8.2 ActSim simulation process

8.3 Chapter Summary

This chapter constructs the location model and develops a location simulation algorithm to simulate the changing process of urban households and business distribution, i.e., the interaction process between transportation and land use. The location simulation algorithm is an integrated application of the aforementioned sub-models, first integrating transportation accessibility and location costs to evaluate location utility, representing the value or attractiveness of zones for various types of households or businesses. Then, based on location utility and the distribution of buildings, it determines the locations of various types of households or businesses. The output of the location simulation algorithm, which is also the output of the ActSim model, predicts the number of households or businesses in each zone for the future. As different applications require different spatial and temporal scales, this chapter discusses the spatial and temporal scales of the model and the algorithm's convergence.

References

Albouy D, Ehrlich G (2016) Housing demand and expenditures: How rising rent levels affect behaviour and costs-of-living over space and time. NBER. http://cba.unl.edu/academic-programs/departments/economics/about/seminar-series/documents/housingexpenditures.pdf. Accessed 14 Mar 2016

Lowry IS (1964) A model of metropolis RM-4035-RC. Rand Corp., Santa Monica

McFadden D (1978) Modelling the choice of residential location. In: Karlquist A et al (eds) Spatial interaction theory and residential location. North Holland, Amsterdam, pp 75–96

Mumtaz B (1995) Meeting the demand for housing, a model for establishing affordability parameters. The Bartlett Development Planning Unit. 14 Mar 2016. www.ucl.ac.uk/silva/bartlett/dpu/publications/dpu-paper-73.

Niu F, Liu W, Feng J (2016) Modeling urban housing price: The perspective of household activity demand. Acta Geogr Sin 71(10):1731–1740

Chapter 9
Application and Discussion of the ActSim Model

After detailing the structure and components (sub-models) of the ActSim model from Chaps. 4–8, this chapter will demonstrate the application of the ActSim model through case studies.

9.1 Case Study Area and Data

Beijing's rapid expansion has led to a series of metropolitan issues such as environmental pollution, resource shortages, and traffic congestion. To optimize Beijing's urban space and promote coordinated development with surrounding cities, the national strategy for the coordinated development of the Beijing-Tianjin-Hebei region was proposed. It involves gradually relocating non-capital core functions from Beijing to surrounding areas. Concurrently, with the launch of a new round of national spatial planning, Beijing is gradually advancing township-scale spatial planning to align with rigid indicators of city and district planning, including population size and building scale. It is foreseeable that the functional dispersion of Beijing will trigger a restructuring of urban space. Therefore, simulating the urban spatial evolution process to scientifically and proactively predict urban spatial development trends and policy impacts is crucial for formulating rational urban spatial policies, implementing national strategies, and promoting sustainable urban development. This study will use Beijing as a case study to apply the ActSim model.

This study focuses on the central urban area and nearby suburban counties, using streets (towns, villages) as research units, totaling 239 Jiedao (zones at township level), as shown in Fig. 9.1. Spatial data used include administrative divisions, road networks (including highways, urban expressways, national roads, provincial roads, county roads, subways, etc.). We overlay various roads to form a network and calculate transportation costs (i.e., shortest time) between streets (towns, villages). The

F. Niu, *Urban Spatial Evolution Simulation*, Urban Sustainability,
https://doi.org/10.1007/978-981-97-3481-8_9

assessment result is a 239 × 239 matrix, representing traffic conditions between zones.

Socio-economic data: The study utilizes data from Beijing's Sixth Population Census, which records the distribution of various types of families across streets

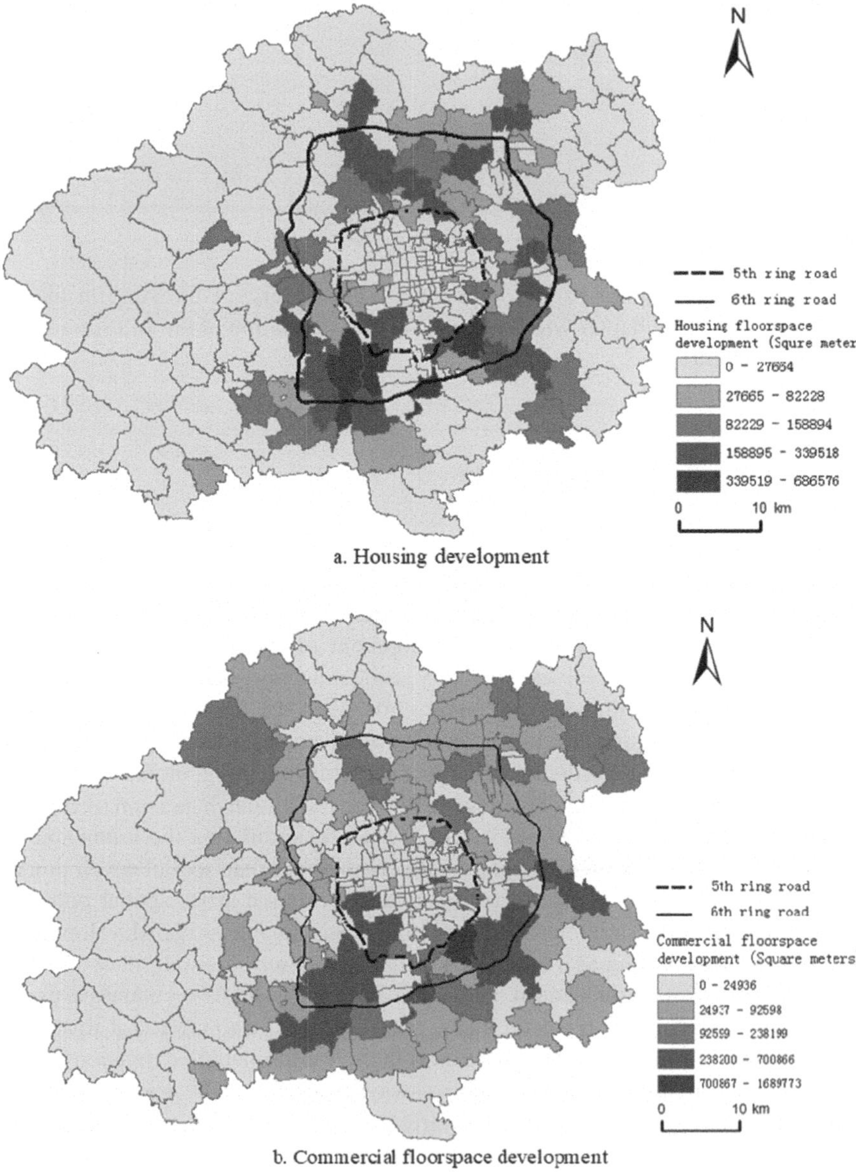

Fig. 9.1 Land use policy. **a** Housing development, **b** commercial housing development

(towns, villages). Spatial location information (POI data) for businesses is obtained through map remote access interfaces (API functions) provided by electronic map suppliers, linking this business data with economic census data to gather enterprise attribute information. For a few unmatched POI data, field research methods are used for supplementation. The business data covers over 700,000 records of companies, schools, research institutes, hospitals, etc., in Beijing.

9.2 Case Application 1: Testing Land Use Policy Scenarios

9.2.1 Policy Scenario Settings

This study first assumes a constant Beijing transportation system and ongoing land use policies in recent years, using the ActSim model to predict the spatial distribution of urban households and businesses.

Corresponding to the urban activity classification, this study divides housing development (land use) into two categories: residential and commercial, measuring development scale by area. The government annually allocates land to developers, specifying the type and maximum area of housing that can be developed on each plot. We compiled land allocation data for Beijing over five years (2009–2013) to calculate the average annual development area for each zone, serving as the basis for future annual housing development, as shown in Fig. 9.1. The figure shows that land use development mainly occurs between the Fifth and Sixth Ring Roads and in suburban county towns, aligning with Beijing's macro policy to alleviate congestion and disperse the central urban population. The distribution of commercial property development is more scattered, leading to further decentralization of employment, primarily in the southern city.

In addition to transportation and land use, the scale of various urban activities also continues to change. We refer to the average growth rate of population and business size in recent years, setting the annual growth rate of urban households and economic activities at 2.3% and 2.0%, respectively.

9.2.2 Forecast Results

Based on the above scenario settings, the ActSim model predicted the distribution of population and economic activities up to 2030. A brief analysis of the forecast results follows:

(1) **Urban Population Distribution Pattern**

By 2030, most of the population will still be concentrated within the Fifth Ring Road, the historic highly developed central urban area, which has accumulated a large population and businesses, offering high accessibility. Despite the continuing inflow of population and the existing base, the central urban area remains the most densely populated.

Figure 9.2b shows the intensity of population growth by 2030 relative to 2010: the percentage change in population for each zone by 2030. The figure indicates that, as the population grows year by year, more and more people tend to move to areas outside the Fourth Ring Road. The ActSim model considers the main factors to be living costs, transportation accessibility, and housing distribution. The central urban area has already been highly developed, with a dense population and less new housing development. At the same time, high rents also hinder further population agglomeration. A large number of housing developments in the suburbs have led to a relative decrease in local rents, and the construction of a large number of commercial properties has attracted a large number of economic activities, further enhancing local transportation accessibility, hence a greater population growth in the suburbs.

From Fig. 9.2b, it can be seen that areas with rapid population growth are mostly located along the Sixth Ring Road, between the Fifth and Sixth Rings. Zones where the growth rate is significantly higher than the surrounding areas include Nan Shao (89), Ma Po (73), Hou Sha Yu (236), Bei Zang Village (92), Chang Yang (97), Liang Xiang (98), and Yi Zhuang (173). According to current land use policies, these areas will gradually develop into urban sub-centers. The distribution of these potential sub-centers aligns with the government's goal of developing a multi-center structure to alleviate traffic congestion. Among them, Yi Zhuang is a sub-center that Daxing District is currently focusing on developing, and according to simulation results, Yi Zhuang will gradually gather more population.

From Fig. 9.2b, it can be concluded that transportation routes have a significant impact on the spatial distribution of residents, with people preferring to live along transportation lines. At the same time, some zones far from the central urban area also have a higher population density, typically located in county towns of the far suburbs, such as Changping City North Street (23) and City South Street (24), Shunyi Guangming Street (77) and Victory Street (78), Fangshan Yingfeng Street (51), etc.

(2) **Urban Economic Activity (Business) Distribution Pattern**

Figure 9.3a shows the distribution pattern of economic activities in 2030. In 2030, most of the economic activities are still distributed in the inner main urban area. Besides the main urban area, suburbs with concentrated economic activities include Yi Zhuang (173), Huang Village (95), Wangjing (140), and Jinding Street (233). In the land use development policy scenario (see Fig. 9.1), these zones have a high volume of housing development, leading to a relative decrease in rents, thereby attracting more companies to move in, and the density of economic activities is significantly higher than surrounding areas.

Figure 9.3b shows the intensity of economic activity growth in 2030. The figure shows that a large number of zones in the central urban area and the southwestern area have a growth percentage less than zero, indicating that the economic activities in these zones have decreased compared to 2010. This is due to the lack of commercial property development in these zones. Meanwhile, other areas with a high volume of

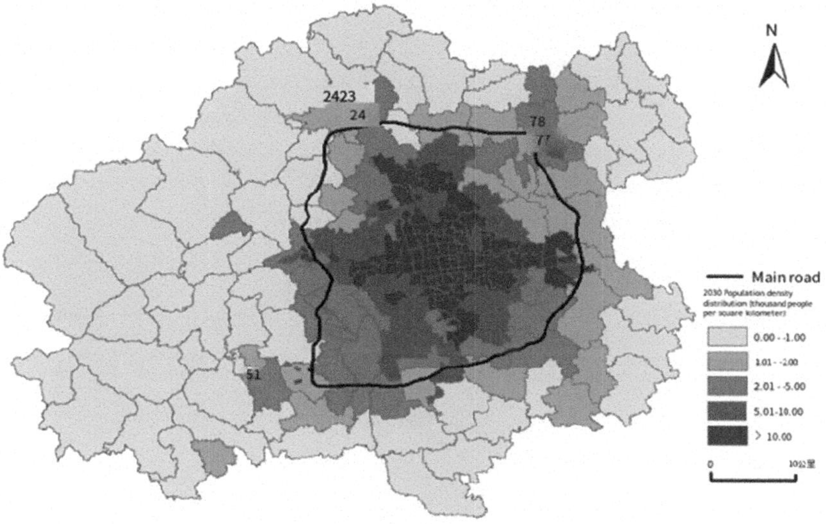

a. Population distribution density

b.Population growth distribution

Fig. 9.2 Forecast of population distribution in 2030. **a** Population density distribution, **b** Population growth distribution

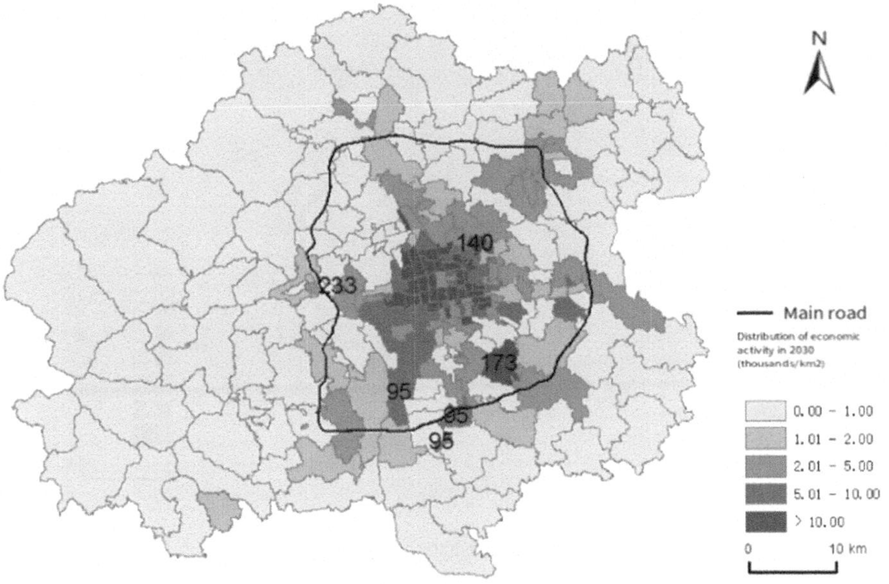

a.Density of economic activity in 2030

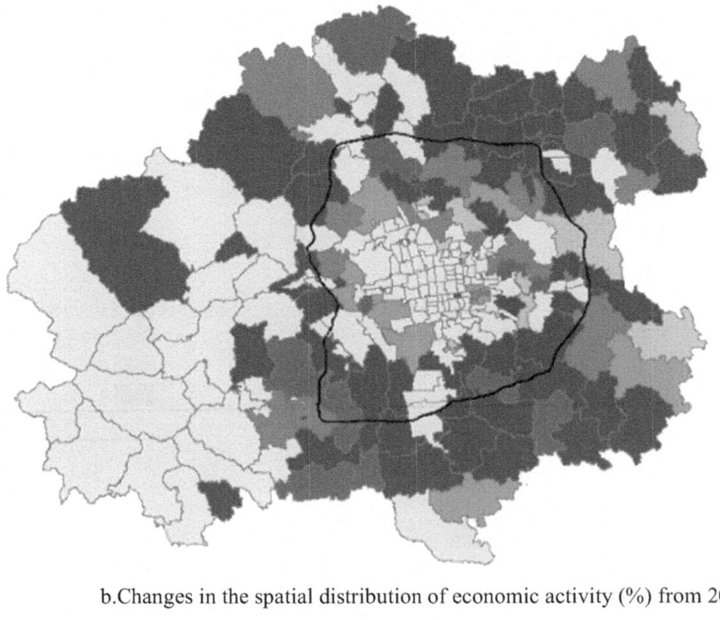

b.Changes in the spatial distribution of economic activity (%) from 2010 to 2030

Fig. 9.3 Projected distribution of economic activity in 2030. **a** Density of economic activity in 2030, **b** change in spatial distribution of economic activity in 2010–2030 (percentage) housing development have attracted more businesses due to the decrease in rents. This simulation result is consistent with the current planning goal of dispersing economic activities from the main urban area, which also validates the model's effectiveness.

Comparing the population growth pattern (Fig. 9.2b) with the economic activity growth pattern (Fig. 9.3b), it can be observed that economic activity growth is more spatially dispersed, with a larger magnitude of growth in the suburbs. This indicates that, driven by the market, business location choices are more sensitive to location costs (rent). With the extensive development of commercial properties in the suburbs leading to a decrease in rents, a significant number of businesses have moved to the suburbs to reduce location costs. It can be anticipated that the relocation of businesses will further drive the migration of the residential population. Additionally, Fig. 9.3b reveals that zones with higher economic activity growth intensity are predominantly located on both sides of the Sixth Ring Road, similar to the commercial land use development pattern (Fig. 9.1b), reflecting that the government can layout urban economic activities through land use policies.

9.3 Case Application 2: Testing Transportation Policy Scenarios

9.3.1 Policy Scenario Settings

The National Land Space Plan for Haidian District from 2017 to 2035 includes the planning of the Xishan Tunnel, which will penetrate the Xishan Mountains, connecting the northern new district of Haidian (including Sujiatuo, Shangzhuang, Wenquan, and Xibeiwang towns) with the central urban area, as shown in Fig. 9.4. It is foreseeable that the opening of this tunnel will significantly enhance the transportation accessibility of the towns in the northern new district of Haidian, attracting more population and businesses. On top of the previous land use policy scenario, we added this tunnel and assumed its opening in 2020 to test its impact on the northern new district of Haidian.

9.3.2 Forecast Results

As shown in Fig. 9.5, after the tunnel's opening, the population and economic activities in the northern areas quickly concentrate and then the growth rate slows down. This is because after the accumulation of population and economic activities, rents inevitably rise, leading to increased location costs and decreased location utility. It can be predicted that as the population and businesses continue to cluster, the locational advantage will further decrease, eventually reaching equilibrium. The results shown in Fig. 9.5 represent the difference from Scenario 1, considered the impact

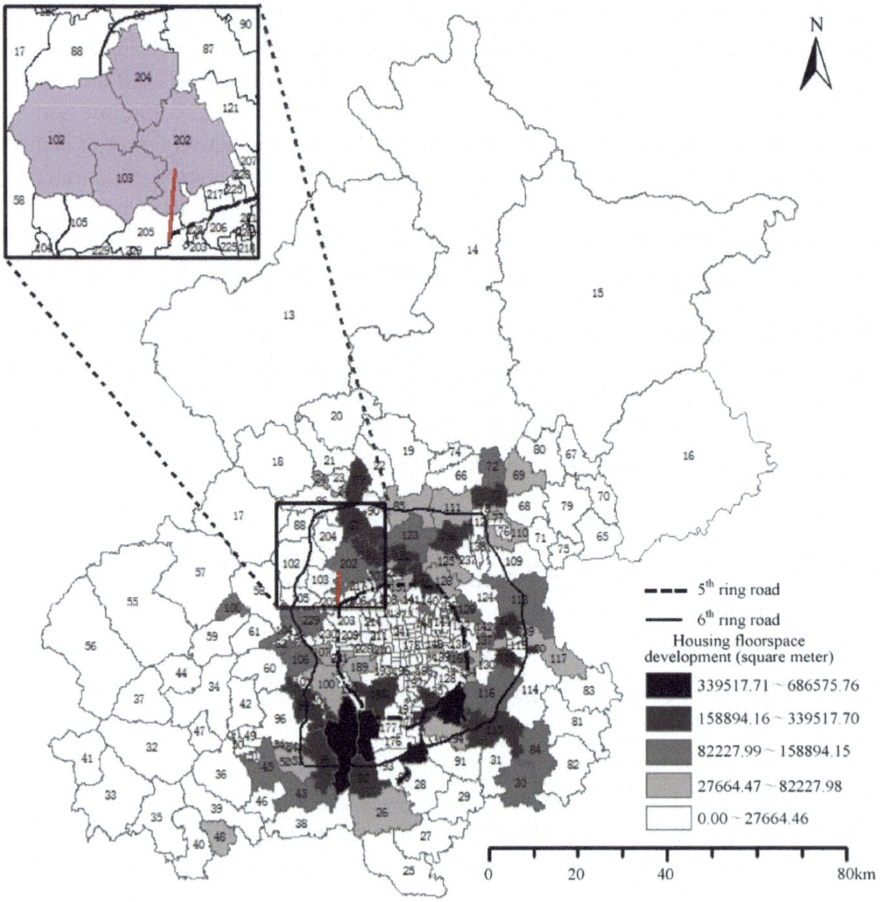

Fig. 9.4 Xishan tunnel

of the tunnel's opening. Focusing on the relative magnitude of simulation results is meaningful, which will be discussed in detail later.

Cities are systems, and as the population and businesses cluster in the Haidian northern new district, other zones will also experience varying degrees of change. Based on this simulation result, the impact of the tunnel on all zones in Beijing can be analyzed. It is noted that the assumption that past land use policies extend into the future in the above case is set to explain the function of the model and does not represent official policy.

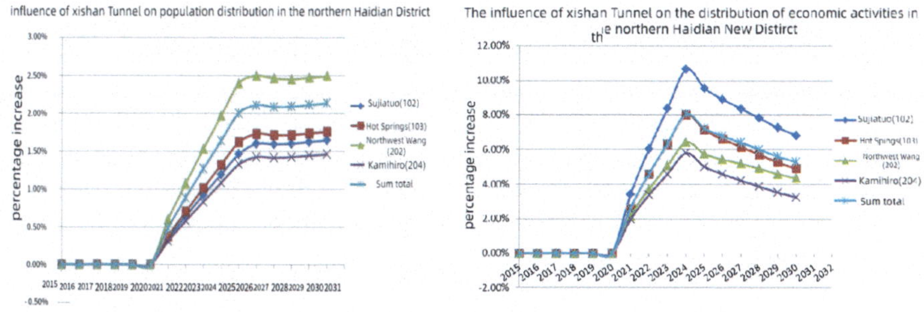

Fig. 9.5 Influence of Xishan tunnel on the northern Haidian new district

9.4 Discussion of the ActSim Model

9.4.1 Model Accuracy and Use

Accuracy is key in modeling. During the calibration process of the ActSim model, we continuously adjust parameters by comparing the predicted values of rents with observed values (actual values), thereby determining the optimal solution for parameters. However, we acknowledge that the ActSim model cannot accurately predict the future of cities, nor do we believe there exists a model that can. Cities are complex mega-systems, influenced by a multitude of factors, including unforeseeable ones. Therefore, in terms of supporting decision-making, the relative value of simulation results is more meaningful than their absolute value. For instance, the difference in simulation results between scenarios with and without a certain policy implementation indicates the impact of that policy.

The LUTI theory is considered to describe the general laws of urban spatial development. Thus, with varying degrees of modification and calibration, the ActSim model can be further applied to other cities. Through more case applications and continuous improvements in variable selection and parameter settings, the model can be progressively refined. The construction and application of this model help promote the application of LUTI models in China, assist in urban spatial decision-making, and enrich the research content of urban simulation analysis in China.

9.4.2 Model Limitations and Improvements

The Lowry model (Lowry 1964) is a milestone in the development of LUTI models. Compared to the initial Lowry model, the ActSim model is dynamic, allowing various elements to change over time to predict future urban spatial conditions. The ActSim model is a balance model. Some views suggest that due to the mismatch between supply and demand, urban spatial development can only tend towards but never reach a state of equilibrium (Kryvobokov et al. 2013; Torrens 2000). Thus, compared to the dynamic process of urban space, the ActSim model is a "quasi-dynamic model."

In case applications, urban location utility evaluation mainly considers transportation accessibility and location costs, but urban activity location choices are influenced by multiple factors, such as family member structure, housing comfort, environment, etc. Incorporating a variety of factors is a direction for future expansion. This task will present researchers with challenges in data acquisition, complicate the model algorithms, and significantly increase computation time.

9.4.3 Optimization and Extension of the Transportation Model

Transportation is a key factor affecting the location of urban socio-economic activities. The ActSim model uses time cost to evaluate transportation conditions, but transportation cost can also be assessed using economic cost or a combined cost of time and economics. Including economic costs means transportation pricing policies will affect the accessibility evaluation results, thereby influencing the model's operational outcomes. Thus, the model can also be expanded to evaluate the impact of transportation pricing policies on urban space.

The transportation model calculates transportation costs using vehicle speeds. However, cities have different transportation modes, such as walking, cycling, public transport, and driving, each with different travel speeds and costs. Therefore, conducting multi-modal transportation simulation analysis is an urgent research topic. The transportation field is highly competitive with many mature transportation software available, such as START, TRAM, etc. The transportation cost evaluation (gc) in the ActSim model can utilize mature transportation software.

9.4.4 Changes in the Total Amount of Urban Activities

This article predicts changes in the total amount of households and economic activity scales based on historical population and business growth rates. However, the actual process of changes in household numbers is complex. The change in the total amount of households is influenced by local household transitions as well as population mobility (in-migration and out-migration). To achieve more accurate predictions of the total number of households, it is necessary to establish a household transition model to simulate local household changes and predict the number of various types of households. Additionally, a population mobility model is needed to simulate population flows between the city and the outside world, considering policy restrictions for simulating external population in-migration.

9.5 Chapter Summary

This chapter conducts case applications of the ActSim model, simulating the urban spatial evolution process in Beijing. Assuming the continuation of recent land development trends, it forecasts the future distribution of urban population and economic activities and tests the impact of the opening of the Xishan Tunnel in Beijing on the Haidian Northern New District based on these scenarios. The study shows that the ActSim model can quantitatively predict the scale of population and economic activities in each zone, providing an excellent tool for testing urban land use and transportation policies. As a demo system, the ActSim model aims to explain the principles, architecture, and implementation of the Land Use-Transport Interaction (LUTI) model to readers. There are many areas where ActSim needs improvement, and this chapter discusses the model's use, accuracy, and limitations. From the model's development perspective, activity-based models align with the LUTI model development trend (Wegener 2004). As an initial exploration of applying LUTI theory to Chinese cities, this study contributes to the development of LUTI models and their application in China, enriching the methods of urban spatial research within the country.

References

Kryvobokov M, Chesneau JB, Bonnafous A et al (2013) Comparison of static and dynamic land use-transport interaction models. Transp Res Rec J Transp Res Board 2344(1):49–58

Lowry IS (1964) A model of metropolis RM-4035-RC. Rand Corp, Santa Monica

Torrens PM (2000) How land-use transportation models work. Centre for Advanced Spatial Analysis, London

Wegener M (2004) Overview of land-use transport models. In: Hensher DA, Button K (eds) Transport geography and spatial systems. Elsevier, Oxford, pp 127–146

Chapter 10
Regional Simulation Framework

The preceding Chaps. 1–9 have focused on elucidating concepts related to urban space and simulation analysis methods, detailing the construction process of Land Use-Transport Interaction (LUTI) models. LUTI models are used to simulate the internal spatial evolution of cities. However, when expanding the research scale to the regional level, which includes multiple cities or urban agglomerations, regional spatial evolution simulation becomes relevant. Although not the main focus of this book, regional simulation is closely linked to urban simulation. Therefore, this chapter makes a preliminary exploration of regional evolution simulation methods, establishing a regional simulation framework to lay the foundation for extending research to the regional level in the future.

Within a region, the spatial differentiation of production and demand generates intercity trade. Additionally, differences in economic development among cities lead to changing population distribution patterns. Correspondingly, intercity connections (interactions) include both trade links and population migration. Hence, regional simulation mainly involves simulating intercity economic connections and intercity population migration, necessitating the creation of a Regional Economic Model (REM) and a Migration Model (MM), as shown in Fig. 10.1.

10.1 Regional Economic Model (REM)

The Regional Economic Model (REM) simulates intercity economic relations. Based on the distribution of production capacity and consumption demand, it determines the trade volume between cities and, based on intercity trade, calculates each city's total output. Intercity trade connections can be divided by different industries. For simplicity, let's assume the region contains X cities and Y industries. For each industry, the result of intercity trade evaluation is a matrix, representing the trade

F. Niu, *Urban Spatial Evolution Simulation*, Urban Sustainability,
https://doi.org/10.1007/978-981-97-3481-8_10

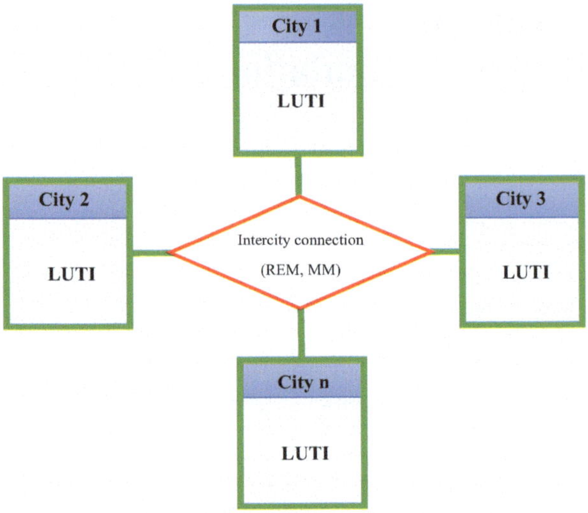

Fig. 10.1 Regional spatial evolution simulation architecture

between every pair of cities. The evaluation of all factors in this model uses a uniform unit—monetary units.

Producers decide their output (or investment) based on estimated trade volumes, which are often assessed based on demand distribution, product prices, transportation costs, and import/export factors. Thus, intercity trade is determined by the distribution of production capacity, demand distribution, product prices, transportation costs, other costs, and import/export influences, where production capacity is affected by investment. REM simulates intercity trade volumes based on these factors and, based on intercity trade volumes, simulates logistics volumes between cities, as shown in Fig. 10.2.

As shown in Fig. 10.2, intercity trade is influenced by various factors such as transportation, capacity, total demand, and product prices. Before evaluating intercity trade, it is necessary to establish sub-models to assess each influencing factor.

10.1.1 Transportation Evaluation—Transportation Model (TM)

Transportation is the infrastructure of intercity connections and a significant factor affecting regional spatial structures. The transportation model first evaluates intercity transportation costs based on the regional transportation network. It then evaluates the locational advantage of each city based on market demand, known as transportation accessibility, reflecting each city's market potential.

Different industries' products have different natures, leading to varying transportation costs. As discussed earlier, transportation cost evaluation can consider time costs, economic costs, or a combination of both, referred to as generalized costs

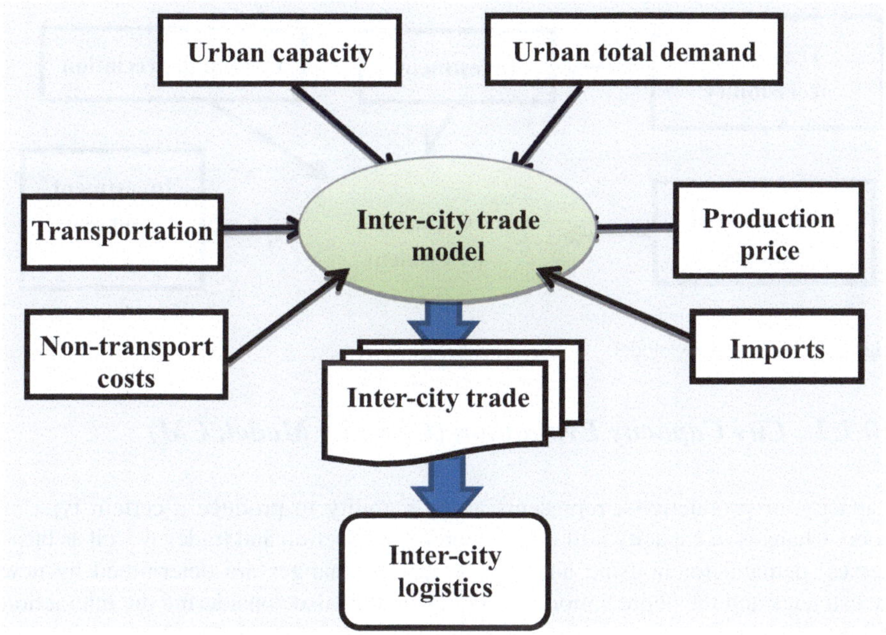

Fig. 10.2 Regional economic model

(Niu et al., 2016). Unlike intra-city transportation, intercity transportation costs are usually evaluated using economic costs. With multiple paths existing between cities, transportation models can use GIS algorithms to calculate the least cost path. For a given industry, the regional transportation cost evaluation result is an X by X matrix, representing the transportation costs between every pair of cities. If there are Y types of industries, the transportation cost evaluation results in Y matrices.

Transportation accessibility reflects each city's locational advantage within the region, influencing investment location choices. A city's transportation accessibility represents the ease of reaching all other cities in the region, considering both the transportation costs to other cities and the opportunities (or market size) those cities offer. Like transportation cost evaluation, transportation accessibility is also assessed by industry and city. For example, for industry Y, city X's transportation accessibility is determined by the regional demand for Y products and the transportation conditions (transportation costs) from city X to other cities. For products that do not require transportation, a different model is needed to evaluate locational advantage without considering transportation costs.

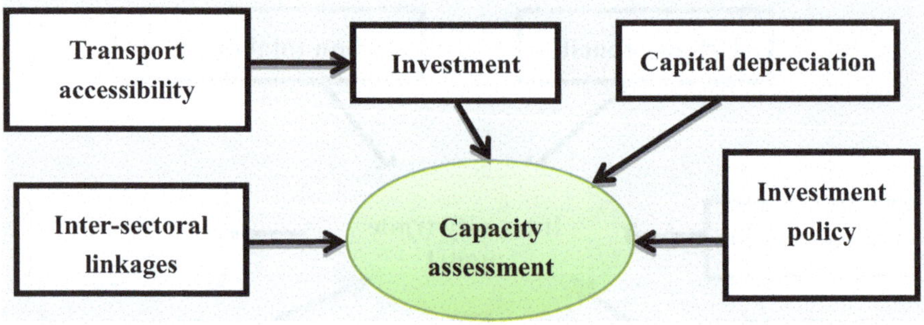

Fig. 10.3 City capacity model

10.1.2 City Capacity Evaluation (Capacity Model, CM)

Capacity, or productivity, represents a city's ability to produce a certain type of good. Changes in capacity will affect intercity production and trade, as well as businesses' demand for housing and land. Capacity changes are determined by new investments and the depreciation of original capital, also considering the interaction among industries, as changes in one industry's capacity can affect the capacities of others.

Investment is a crucial factor affecting capacity. To forecast capacity distribution, it is first necessary to predict the investment distribution pattern. City investments are influenced by existing capacity, market potential, transportation, and various costs. Capital depreciation refers to the capacity left from the previous year after depreciation. Additionally, some industries are interlinked with others, so capacity evaluation must consider changes in other related sectors to determine the impact of industry interconnections. Based on the evaluation of these elements, further modeling can evaluate city capacity (by industry), as shown in Fig. 10.3.

10.1.3 City Total Demand Evaluation (Demand Model, DM)

Total demand refers to each city's total demand for various goods, a crucial factor influencing trade patterns. Total demand includes both final demand and intermediate demand. Final demand primarily consists of government consumption, consumer final demand, exports, and total capital formation, among others. The processing and distribution processes required to meet the city's final demand generate intermediate demand for other goods. Intermediate demand can be evaluated based on final demand and inter-industry input–output coefficients (I/O coefficients), further evaluating total demand based on final and intermediate demand, as shown in Fig. 10.4. Total demand evaluation is also conducted by city and industry.

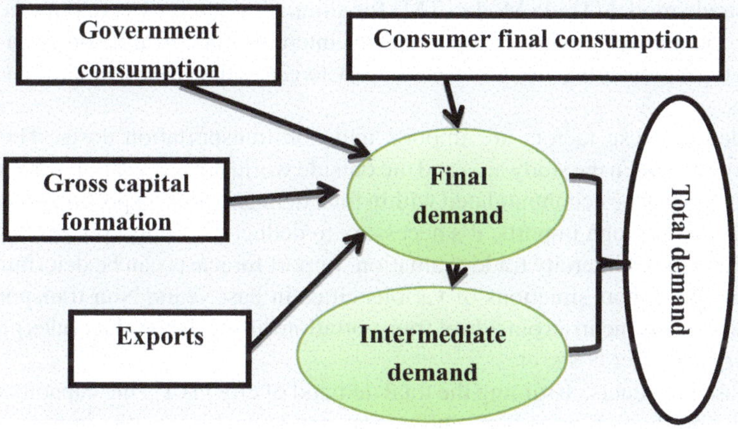

Fig. 10.4 Demand model

10.1.4 Product Price Evaluation (Price Model, PM)

Product price refers to the production price per unit of product, determined by raw material costs, location costs, labor costs, added value, and other factors. Price calculation is performed by industry, representing the average price of all goods within an industry. Raw material cost refers to the consumption of various other products required to produce a particular type of product, evaluated based on input–output relationships, involving other goods' prices, input–output coefficients, and also considering transportation costs. The evaluation of transportation costs needs to consider the distribution of raw materials in other cities, the likelihood of using raw materials from each city, and then collectively evaluate the total transportation cost based on the aforementioned transportation cost evaluation results. Location cost refers to expenses arising from leasing factories or workspace, and some industries may have significant non-rental location costs, such as water, electricity, property fees, etc. Labor costs can be calculated using the compensation for hired workers. Added value refers to the value added generated during the production and sales process of various products. The price model needs to integrate all these costs to calculate product prices.

10.1.5 Simulation of Intercity Economic Connections

The evaluation of economic connections aims to determine the trade volume between pairs of cities, that is, the total amount of various products a city sells to another. This evaluation is based on the results of transportation, capacity, total demand, product prices, along with factors like imports and non-transportation costs, to construct an intercity trade model (Trade Model, TM) for simulating intercity economic connections. As mentioned previously, the result of intercity trade evaluation is an $Y * B * B$ matrix, representing the trade volume between pairs of cities across various industries.

Included in these factors are imports and non-transportation costs. The trade connections between the study area and the outside world are termed imports/exports, where exports can be accommodated within final demand. Since a portion of the total demand is met through imports, it's necessary to deduct the import portion from the total demand in the intercity trade evaluation. Import forecasts can be determined by examining the import situations of various cities in past years. Non-transportation costs refer to costs incurred outside of transportation during the product sales process, such as cross-border costs, etc.

For class s products, assuming the total demand of city j is Y_j^s, the capacity of city i is K_i^s, the production price is p_j^s, the transportation cost is c_{Tij}^s, the import volume of city j is I_j^s, and other costs are b_{ij}^s, then the total volume T_{ij}^s of class s products sold from city i to city j is a function of the aforementioned variables, as shown in Eq. 10.1. This model comes with corresponding constraints, namely that the sum of goods from various cities to city j cannot exceed city j's total demand.

$$T_{ij}^s = f\left(Y_j^s, I_j^s, K_i^s, p_j^s, c_{Tij}^s, b_{ij}^s\right) \tag{10.1}$$

The city's total output can be determined based on the city's net trade output. Actual scenarios may involve production surplus or other unusual situations, which REM does not consider for the time being. Additionally, a city may have a resistant attitude towards goods from another city; for such cases, variables can be introduced to reduce the simulated value of trade. Within the same region, similar situations can usually be disregarded.

10.1.6 Intercity Freight Model (Freight Model, FM)

Intercity trade ultimately translates into logistics. Goods that do not require transportation are not considered for transport. Logistics distribution will classify and transport products based on their characteristics. When the packaging and transportation methods of products differ, the volume of each transportation will also vary. The intercity transportation volume (number of freight trips) can be calculated based on the volume of each transportation and the total volume of intercity trade. Converting trade volume to logistics volume requires research into the logistics industry, combined with the actual business of various logistics companies to build the model. It is difficult to accurately predict logistics volume, but the model can ensure the relative size of forecast values between different cities is reasonable. Thus, this model can play its role in the analysis of regional spatial structures.

10.2 Migration Model (Migration Model, MM)

In addition to economic connections, intercity connections also include population migration. Here, "population migration" refers to the relocation activities of the population, not instantaneous population movements (such as tourism, business trips, etc.), which will change the total population of cities and the regional population distribution pattern. In 1966, Lee proposed the "push–pull" theoretical framework for population migration, i.e., intercity population migration is determined by the push factors of the origin, the pull factors (or attraction) of the destination, distance impedance, and also influenced by the characteristics of the population itself (Lee 1966). Urban population migration includes various groups; if only this macroscopic population migration is simulated, specific population characteristics may not be considered. Accordingly, this paper presents an implementation framework for the "push–pull" population migration model. As shown in Fig. 10.5, the implementation of the intercity population migration model includes the evaluation of push factors from the outflow city, the evaluation of pull factors from the inflow city, and the evaluation of distance impedance, among others.

10.2.1 Evaluation of City Push and Pull Forces

Before evaluating city push and pull forces, it's necessary to determine the factors influencing population migration. The same factor can be both a push factor and a pull factor; for example, environmental quality can attract population, while poor environment can also hinder population inflow. Common factors affecting household relocation include city size (S), economic opportunities (such as job opportunities) (J), housing costs (R), environmental quality (E), wage levels (I), etc. Actual modeling

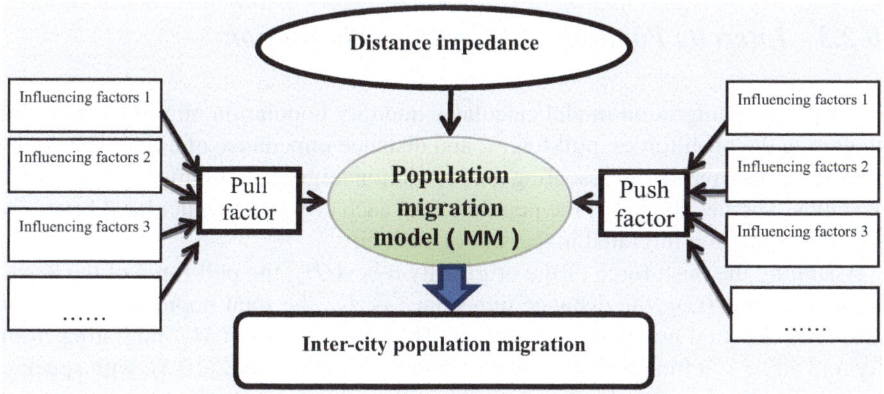

Fig. 10.5 Migration model

requires determining the influencing factors and weights based on the specific situation of the case area, and on this basis, establishing push and pull force evaluation models, as shown in Eq. (10.2). Each city has a push value and a pull value. The evaluation of push and pull forces should use the same indicators (i.e., factors influencing population movement). Some influencing factors are positively correlated with city push forces and negatively with city pull forces, like housing prices, where higher values increase city push forces and decrease pull forces; conversely, factors like income, where higher values decrease push forces and increase pull forces. The modeling process needs to address the positive and negative issues of variables.

$$v(O/D) = f(S, J, R, E, I) \tag{10.2}$$

10.2.2 Distance Impedance Evaluation

Distance is an important factor that acts as an impediment to population migration. The greater the distance, the greater its impedance effect. Distance impedance evaluation determines the variation of distance impedance with distance. Evaluating intercity distance impedance requires determining the intercity distance, usually measured using transportation cost, derived from the output of the transportation model. Since different migrating populations have different sensitivities to distance, such as varying by gender, age, education, the impact of migration distance will differ. For precise evaluation, it's necessary to set different parameters for different populations to categorize distance impedance evaluation. If not classifying the migrating population but simulating the overall population migration, a unified evaluation parameter can be set.

10.2.3 Intercity Population Migration Simulation

The population migration model calculates intercity population migration flows by integrating the push forces, pull forces, and distance impedance of cities. The simulation result is a matrix, representing the population migration volume between every two cities. The net change in the population of each city can be calculated based on the total population migrated in and out of each city.

Assuming the push force of the origin city a is $v(O)_a$, the pull force of the destination city z is $v(O)_z$, the distance impedance is d_{az}, the total population of city a is H_a, and the total population of city z is H_z. The population M_{az} migrating from city a to city z is a function of these variables, as shown in Eq. (10.3), with specific implementation details to be determined based on this framework.

$$M_{az} = f(H_a, v(O)_a, d_{az}, H_z, v(O)_z) \tag{10.3}$$

In applications, it's necessary to ensure the population migration simulation values are within a reasonable range. For example, the population migrating out of a city cannot exceed the total population of that city; the total population migrating into a city cannot exceed policy limits.

10.2.4 Simulation of Population Migration Between the Region and the External World

Each city within the region experiences population migration with the external world (areas outside the region), including both in-migration and out-migration. Population migration between the region and the external world is influenced by the socioeconomic development within and outside the region. The model's implementation can be divided into two steps: first, simulating the total population migration (in-migration and out-migration) between the entire region and the external world, and then allocating the total population migration to each city, determining the population migration volume of each city with the external world. Additionally, changes in city populations also involve natural population growth. This chapter focuses on intercity economic connections and does not discuss this further.

10.2.5 The Role of the Model

The regional model serves two main purposes. Firstly, it can be used to analyze intercity connections over the years, calculating past intercity connections based on historical data. These calculations provide foundational data for compiling intercity input–output tables. Secondly, it is used to assess the impact of regional spatial policies. For example, the development of a highway within a city cluster will change the regional transportation system—altering intercity transportation costs—affecting urban accessibility, impacting regional investment patterns, affecting urban capacity distribution, and ultimately changing the regional trade patterns. Changes in the investment pattern will also alter the distribution of job opportunities and affect population migration. This series of logical relationships is reflected in the aforementioned models. In summary, once the model is constructed, it can be flexibly applied to assess the impact of various policies.

10.3 Chapter Summary

Previous chapters have focused on explaining urban spatial simulation analysis methods and detailed the construction process of the Land Use-Transport Interaction (LUTI) model (ActSim) for simulating the internal spatial evolution of cities. This chapter has established a regional model framework, including a regional economic model and a population migration model for simulating intercity trade and population migration, respectively. The construction of the regional model framework lays the foundation for extending research to the regional level. The implementation of the regional model framework, together with the urban model described earlier, forms a model system.

The output of the regional model is matrices of intercity trade and population migration. Among them, the intercity trade matrix provides basic data for compiling input–output tables. Currently, the compilation of input–output tables is mainly at the provincial level (Liu et al. 2012), and it depends on related intercity connection data, which is often difficult to obtain. The development of this model will provide basic information for compiling input–output tables at different scales (provincial, municipal, county). The regional model framework constructed in this chapter includes many sub-models. The implementation of the model can be refined to different levels according to specific application needs.

References

Lee ES (1966) A theory of migration. Demography 3(1):47–57

Liu W, Liu H, Fan X et al (2012) Sector-specific spatial statistic model for estimating inter-regional trade flows: A case study of agriculture,chemical and electronic sectors in China. Acta Geogr Sin 67(2):147–156

Niu F, Liu W, Feng J (2016) Modeling urban housing price: The perspective of household activity demand. Acta Geogr Sin 71(10):1731–1740

Chapter 11
Common Tools and a Brief Analysis of Mathematical Models

As research advances and information technology develops, an increasing number of analytical and simulation tools have emerged, facilitating research in the academic community. Although various analytical tools can collectively be referred to as software, there are conceptual differences among development tools (languages), simulation tools, and analysis software. This article introduces and distinguishes common tools, as well as introduces commonly used models and functions.

11.1 Development Languages

11.1.1 MatLab

MatLab, along with Mathematica and Maple, is known as one of the three major mathematical software packages, or more accurately, a development language. In the realm of mathematical and technological application software, MatLab leads in numerical computation. MatLab, a combination of "Matrix" and "Laboratory," is released by the American company Mathworks, primarily targeting scientific computing, visualization, and interactive program design. Integrating powerful functions such as numerical analysis, matrix computation, scientific data visualization, and modeling and simulation of nonlinear dynamic systems into an easy-to-use window environment, MatLab offers a comprehensive solution for scientific research, engineering design, and all scientific domains requiring numerical calculations. It significantly deviates from the traditional non-interactive programming languages (e.g., C, Fortran), representing the advanced level of today's international scientific computing software.

MatLab's product series can be used for numerical analysis, numerical and symbolic computing, engineering and scientific drawing, control system design

F. Niu, *Urban Spatial Evolution Simulation*, Urban Sustainability,
https://doi.org/10.1007/978-981-97-3481-8_11

and simulation, digital image processing technology, digital signal processing technology, applications in communication system design and simulation, financial engineering, management, and optimization calculation (operations research). The basic data unit in MatLab is the matrix, making matrix problem solving simpler. MatLab also incorporates advantages from software like Maple, enhancing its capabilities as a powerful mathematical tool. The latest versions have added support for C, C++, and Java.

11.1.2 Mathematica

Mathematica offers a user-friendly development interface and strong mathematical computing capabilities, combining numerical and symbolic computing engines, a graphics system, a programming language, and a documentation system, along with connectivity to other applications. Many of its functions lead globally, especially in symbolic computation. Mathematica is one of the most widely used mathematical software, a powerful system in the general computing field. Since its release in 1988, Mathematica has profoundly impacted how computers are utilized in technology and other fields, marking the beginning of modern scientific computing.

11.1.3 Maple

Maple, renowned in the fields of mathematics and science as "the mathematician's software," is one of the most universal mathematical and engineering computing softwares worldwide. With millions of users globally, it is extensively used in science, engineering, and education, covering over 96% of the world's major universities, research institutions and over 81% of the Fortune 500 companies.

The Maple system includes advanced modeling and simulation of mathematical problems, such as symbolic computation, infinite precision numerical computing, innovative internet connectivity, and a powerful 4GL language. With over 5000 computational commands and mathematical and analytical functions covering nearly all mathematical branches, Maple not only provides programming tools but also mathematical knowledge. It is an essential scientific computing tool for professors, researchers, scientists, engineers, and students, capable of solving simple numerical computations to complex nonlinear problems efficiently. Users can complete multi-disciplinary physical system modeling and simulation, symbolic computation, numerical computation, programming, technical documentation, report presentations, algorithm development, and external program connections in a single environment with Maple products, meeting the needs of users from high school students to advanced researchers.

11.1.4 Other Common Languages

R is a language and environment created for statistical computing and graphics, mainly used for statistical analysis, drawing, and data mining (Long 2021). R is an open-source data analysis solution, running on various systems like Windows, UNIX, and Mac OS X. As a comprehensive statistical research platform, R offers a wide range of data analysis techniques, encompassing almost any type of data analysis, including advanced statistical computing routines not available in other software. Besides its robust modules and built-in functions, R can extend existing methods through custom functions, offering easy expandability, high flexibility, and rapid updates. Additionally, R emphasizes interactive data analysis, allowing easy data import from various sources, such as text files, database management systems, statistical software, and even specialized data warehouses, as well as direct data retrieval from websites, social media sites, and various online data services. It also interfaces well with other programming languages and databases. Moreover, R boasts top-tier charting capabilities, offering significant convenience for those visualizing complex data (Bivand et al. 2013). In summary, for statistical data analysis and drawing, R is a comprehensive and ideal development tool (Kabacoff 2016).

Python is an interpreted, object-oriented, and dynamically typed high-level programming language, widely used for its simplicity, readability, and extensibility, along with its rich library, making it one of the most popular programming languages. Compared to languages like C++ and JAVA, Python's style and data types are concise, with less code and frequent use of English keywords and Tab indentation to regulate structure, offering high readability and easier learning. If proprietary algorithms need to be written, other languages can complete that part of the program, which can then be called from Python. Additionally, the Python standard library provides a wealth of functionality, including text/binary data handling, mathematical operations, functional programming, file/directory access, data persistence, data compression/archiving, encryption, operating system services, concurrent programming, inter-process communication, network protocols, JSON/XML/other Internet data formats, multimedia, GUI, debugging, profiling, etc. Python also interfaces with all major commercial databases, creating significant convenience for users and enjoying widespread popularity (Mckinney 2018). The Python language supports powerful functionality for urban spatial data analysis and operational simulation.

11.2 Analysis Software

11.2.1 Excel

Microsoft Excel is a spreadsheet software developed by Microsoft for computers running Windows and Apple Macintosh operating systems, a crucial part of the Microsoft Office suite. Its intuitive interface, excellent calculation, and charting tools, coupled with successful marketing, have made Excel the most popular personal computer data processing software (Microsoft Excel 2018). Excel's main functions include data recording and organization, data processing and calculation, data statistics and analysis, graphic report production, information transmission, and sharing, widely applied in management, statistical finance, finance, and many other fields. Besides its powerful data processing capabilities, Excel also incorporates the VBA language, allowing users to customize functionality and develop individualized automation solutions.

11.2.2 ArcGIS

ArcGIS products, developed by ESRI, offer users a scalable and comprehensive GIS platform for spatial data analysis and visualization. ArcGIS is based on a set of common component libraries called ArcObjects, which contain a large number of programmable components, integrating comprehensive GIS functionality for developers.

As a scalable platform, ArcGIS provides GIS functionality to individual and group users through desktop, server, web, and other means. ArcGIS 10 is a complete collection of GIS software, including frameworks for deploying GIS: ArcGIS Desktop as a complete professional GIS application suite; ArcGIS Engine as embedded development components for customizing GIS applications; server GIS includes ArcSDE, ArcIMS, and ArcGIS Server, a set of components for developing B/S or C/S systems.

Researchers most commonly use desktop GIS (ArcGIS Desktop). ArcGIS Desktop is a software suite integrating numerous advanced GIS applications, including a set of Windows desktop applications with user interface components (e.g., ArcMap, ArcCatalog, ArcToolbox, and ArcGlobe). With its powerful spatial analysis and visualization capabilities, ArcGIS is a common and even essential tool in the geographic community.

11.2.3 SPSS

SPSS, along with SAS and BMDP, is one of the three most influential statistical software internationally, incorporating data entry, organization, analysis, and output management. SPSS's basic functionalities include data management, statistical analysis, chart analysis, output management, etc. Its statistical analysis processes cover descriptive statistics, mean comparison, general linear models, correlation analysis, regression analysis, log-linear models, cluster analysis, data reduction, survival analysis, time series analysis, multiple responses, and more, with several statistical processes under each category, such as linear regression analysis, curve estimation, logistic regression, probit regression, weighted estimation, two-stage least squares, nonlinear regression, and others. SPSS also has a dedicated plotting system for creating various charts based on data.

SPSS can directly read Excel and DBF data files, offering clear, intuitive, and easy-to-learn and use analysis results. SPSS has been extended to various operating systems. Users can choose modules according to their actual needs and computer capabilities to reduce system hard disk requirements, effectively promoting the software's dissemination and application.

11.2.4 GeoDa

GeoDa is a software toolkit for exploratory spatial data analysis (ESDA) of raster data. It provides users with a user-friendly graphical interface for spatial data analysis, such as autocorrelation statistics and outlier indicators. GeoDa features an operating environment combining maps and statistical charts. Initially designed to bridge between ESRI's ArcInfo GIS and SpacStat software for spatial data analysis, the current GeoDa is a standalone software that does not require a specific GIS environment. GeoDa can run on all Microsoft operating systems (Win95, 98, 2000, NT, Me, Xp), and its installation system includes all necessary files.

11.2.5 ENVI

ENVI (The Environment for Visualizing Images) is the flagship product of Exelis Visual Information Solutions, USA. Developed using the Interactive Data Language (IDL) by scientists in the field of remote sensing, ENVI is a powerful software for remote sensing image processing, enabling fast, convenient, and accurate information extraction from images. ENVI supports images from various types of aerial and spaceborne sensors, including panchromatic, multispectral, hyperspectral, radar, thermal infrared, terrain data, GPS data, and LIDAR, among others. Capable of reading over 80 data formats, including HDF, Geodatabase, GeoTIFF, and JITC-certified NITF formats, ENVI's enterprise-level performance allows for rapid access to images on OGC and JPIP compatible servers through internal organizational structures or the internet. Today, numerous image analysts and scientists choose ENVI for extracting information from remote sensing imagery. ENVI is widely applied in fields such as scientific research, environmental protection, meteorology, oil and mineral exploration, agriculture, forestry, medicine, defense, earth sciences, public utility management, remote sensing engineering, water resources, oceanography, surveying, and urban and regional planning.

11.2.6 AMOS

AMOS is a software used for exploring relationships between variables through structural equation modeling (SEM). It facilitates easy SEM modeling, enabling quick model creation to test the interactions and causes among variables, providing more accurate and comprehensive analysis results than solely using factor analysis or regression analysis. Structural equation modeling is a multivariate analysis technique that includes standard methods and extends them. These methods encompass regression analysis, factor analysis, analysis of variance, and correlation analysis. AMOS makes SEM analysis straightforward with its intuitive drag-and-drop drawing tool, allowing for quick customization of models without programming. Every step in building a structural equation model with AMOS provides a graphical environment, where models can be specified or altered with a click of the mouse. AMOS offers a suite of basic analysis methods, including variance and covariance analysis, hypothesis testing, and also supports Bayesian analysis, bootstrapping, growth curve models, and non-recursive models. Higher versions of AMOS have enhanced transparency and extensibility, providing interfaces for linking with VB, SAS, and other software, greatly facilitating programming.

11.2.7 Ucinet

Ucinet is a network analysis software with a menu-driven Windows program. It integrates with data software NetDraw, 3D analysis software Mage, and large network analysis software Pajek. Ucinet can read files in formats such as text, KrackPlot, Pajek, Negopy, and VNA, and can handle up to 32,767 network nodes. It has strong matrix analysis capabilities, such as matrix algebra and multivariate statistical analysis, offering a wealth of data management and transformation tools. Ucinet is currently the most popular and user-friendly social network analysis software.

11.3 Simulation Software

Simulation software is used for modeling, allowing users to create models, set data, and then run simulations to obtain results. Such software usually provides a what-you-see-is-what-you-get interface, making it user-friendly, and also offers scripting interfaces to enhance simulation capabilities. Essentially, this type of software represents a highly abstract model framework, such as UrbanSim, which is fundamentally a model framework for urban land use-transport interaction. Using such simulation software requires users to have a clear understanding of the principles behind the model. The professionalism and complexity of the model limit the widespread application of this type of software.

11.3.1 Vensim

Vensim offers a simple and flexible way to build system dynamics models, including causal loops, stocks, and flow diagrams. In the Vensim modeling process, connecting variable symbols with graphical arrow notations and entering relationships between variables as equations in the model documents the causal relationships. This process helps understand the causality and feedback loops between variables and the relationship between inputs and outputs, facilitating users to grasp the model structure and allowing modelers to modify the content. Through Vensim's user manual, one can quickly learn and master the use of Vensim by building various simple models.

11.3.2 Netlogo

Netlogo is a programmable modeling environment for simulating natural and social phenomena, proposed by Uri Wilensky in 1999 and developed by the Center for Connected Learning and Computer-Based Modeling (CCL). As a common platform for building agent-based models, Netlogo's world consists of agents with different functions, including patches, turtles, links, and observers. Modelers can instruct numerous independently operating agents through programming, allowing macro patterns to emerge through interactions. Netlogo is particularly suitable for modeling complex systems evolving over time. It comes with a model library covering research in biology, medicine, physics, chemistry, mathematics, and computer science. Netlogo, an upgraded version of the Logo language, overcomes the limitation of controlling only a single entity in Logo, enabling modeling control over thousands of individuals, thus effectively simulating the behavior of micro-level agents and the emergence of macro-level patterns and their interconnections. Netlogo is especially suited for simulating complex systems over time.

11.3.3 UrbanSim

UrbanSim is a city spatial development simulation software developed under the leadership of Dr. Waddell from the University of California. Based on the principles of land use-transport interaction, the software simulates the impact of urban spatial policies on transportation, economic activities, and population distribution. Widely used in the United States, UrbanSim supports the testing and forecasting of urban spatial policies and development. Employing a dynamic non-equilibrium approach to simulate urban spatial development, UrbanSim's core is the land use-transport interaction model (LUTI), requiring a thorough understanding of LUTI principles to effectively use the software. Its complexity makes it a professional tool, which also hinders its widespread adoption.

11.3.4 MEPLAN

MEPLAN software, developed by Professor Marcial Echenique and his colleagues from the University of Cambridge, is a city spatial development simulation software that quantifies traffic volume and social products through trade price adjustments to balance transportation and land use. The software consists of three sub-modules: land use, transportation, and production and consumption. The land use submodule handles goods, services, and labor in a spatially disaggregated manner, using input–output techniques to calculate travel demand. Currently, MEPLAN is primarily used by the UK government.

Similar city spatial development simulation software includes TRANUS, DELTA, PECAS, and ILUTE. These software tools share the LUTI model core but differ in implementation techniques. Further details on each model's implementation can be explored in relevant literature, which is not elaborated here.

11.4 Brief Analysis of Commonly Used Models

11.4.1 Coordination Analysis Model

The coordination degree analysis model is used to analyze the degree of coordination between variables. For example, to determine the degree of coordination between regional infrastructure and the level of socio-economic development, the commonly used formula is

$$C_{xy} = \frac{X + Y}{\sqrt{X^2 + Y^2}} \tag{11.1}$$

where X is an indicator or variable in the study area, Y is another indicator or variable in the study area, and C_{xy} represents the coordination index of indicators X and Y in the study area. C_{xy} has the following properties: ① The size of Cxy is positively correlated with the coordination of the system development, the larger the Cxy, the higher the coordination of the system, and vice versa, the lower the coordination; ② $-1.414 \leq C_{xy} \leq 1.414$, when X, Y are positive and equal, the value of C_{xy} is the largest, 1.414; if X, Y are negative and equal, the value of C_{xy} is the smallest, and the other cases are between the two (Wu et al. 2009).

11.4.2 Discrete Choice Models

Discrete Choice Models (DCM), also known as Qualitative Response Models (QRM) or Choice-Based Conjoint Analysis (CBC), emerged in the late 1950s within the realm of microeconometrics and have found widespread application across various social science fields (Nie and Jia 2005; Yan 2018; Shi 2015). These models are typically employed to study the factors influencing individual preferences under certain contexts. For instance, when deciding on a mode of transportation, people might choose among several options due to differences in time efficiency, cost, convenience, and comfort. Let's denote choosing a bus as $Y = 1$, a taxi as $Y = 2$, walking as $Y = 3$, and driving as $Y = 4$. In studying people's choice of transportation, since the dependent variable is not continuous but categorical, traditional linear regression models have significant limitations, whereas discrete choice models offer an effective modeling approach.

1. **General Principles and Application Value of Discrete Choice Models**

The general principle of discrete choice models is based on Random Utility Theory: it is assumed that a decision-maker has J options, each with a certain utility U, composed of both fixed and random utilities. The fixed utility V can be explained by certain observable variables x, known only to the decision-maker and unknown to the researcher. The random utility ε represents the unobserved utility and error effects. The decision-maker's strategy is to choose the option with the highest utility, hence the probability of each option being chosen can be expressed as a function of its observable utility: $P = F(V)$, with the specific form depending on the distribution of the random utility. In most model settings, the observable utility V is expressed as a linear combination of observable variables x, i.e., $V = \beta x$, where β is a coefficient, its value and significance level can be calculated from observed data.

Taking the choice of transportation mode by residents as an example, the influencing factors include the attributes of the decision-maker and the differences between modes of transportation. The attributes of the decision-maker include income, health status, gender, etc., while modes of transportation include bus, car, cycling, etc., with differences in comfort, cost, safety, etc. Decision-makers will consider all these factors to make a choice that maximizes utility.

Suppose residents choose between two transportation modes i and j, with differences in transit time, cost, comfort, and safety. The observable utility of mode i is represented as:

$$V_i = \beta_{cost} \cdot Cost_i + \beta_{amenity} \cdot Amenity_i + \beta_{security} \cdot Security_i \tag{11.2}$$

Therefore, the utility of mode i is:

$$U_i = V_i + \varepsilon_i \tag{11.3}$$

The observable utility of mode j is represented as:

$$V_j = \beta_{cost} \cdot Cost_j + \beta_{amenity} \cdot Amenity_j + \beta_{security} \cdot Security_j \tag{11.4}$$

Therefore, the utility of mode *j* is:

$$U_j = V_j + \varepsilon_j \tag{11.5}$$

where, the coefficient β represents the weight of each attribute of alternative modes, which needs to be estimated through modeling as a parameter vector. Variables Cost, Amenity and Security represent transportation cost, comfort, and safety, respectively. Since different decision-makers assign different weights to the attributes of alternative options, the impact of the decision-maker's attributes on the decision is reflected in the differences in the coefficient β. In application, residents can be categorized, and the same weight can be used to evaluate the utility for similar residents.

The "utility maximization criterion" means choosing the option with the greatest utility; if $Ui < Uj$, then mode *j* is chosen for travel. Accordingly, the probability of a decision-maker choosing option *i* over *j* can be expressed as:

$$\begin{aligned} P_{ni} &= \text{Prob}\left(U_{ni} > U_{nj}, \forall j \neq i\right) \\ &= \text{Prob}\left(V_{ni} + \varepsilon_{ni} > V_{nj} + \varepsilon_{nj}, \forall j \neq i\right) \\ &= \text{Prob}\left(\varepsilon_{ni} - \varepsilon_{nj} > V_{nj} - V_{ni}, \forall j \neq i\right) \end{aligned} \tag{11.6}$$

Discrete choice models are now widely applied, mainly in market and transportation research. In market research, by analyzing consumers' preferences for different goods and services, they measure, test, and predict market demand. In transportation, discrete choice models are used to analyze individuals' choices of destination, mode of transportation, and route, thereby predicting traffic demand, marking a research frontier. Furthermore, applications of discrete choice models in geography, environmental science, social science, spatial analysis, economics, medicine, education, psychology, and more are also quite common (Wang et al. 2015).

The main value of discrete choice models is manifested in three aspects: First, revealing behavioral patterns through the analysis of the signs, magnitudes, and significance of coefficients β, it is possible to determine which factors genuinely affect behavior and their direction and importance. For different demographic groups, it is also possible to compare differences among groups. Second, estimating willingness to pay, where in market applications, the monetary value of an attribute can be calculated by the ratio of other attributes' coefficients to the price coefficient. This method can also be extended to two non-price attributes. Third, simulation analysis, typically done in a "What-if" manner to examine differences caused by changes in factors, policy implementations, and the addition or removal of options.

2. **Common Discrete Choice Models - The Logit Model**

The logit model, derived from the term "log-it," represents the logarithm of "it," where "it" refers to the odds, which is the ratio of the probability of an event occurring to the probability of it not occurring, mathematically expressed as:

$$odds = \frac{P_i}{1 - P_i} \tag{11.7}$$

The value of P ranges from [0, 1], and the range of odds is $[0, +\infty)$. Taking the natural logarithm of odds maps the range of probability P from [0, 1] to $(-\infty, +\infty)$. This important characteristic of logit, having no upper or lower bounds, brings significant convenience to modeling, as the dependent variable and logit share the same value interval. Assuming X is a sequence of varaibles that affects one's decision and P_{ni} is the probility of people n choose option i, then we can establish the relationship between X and P_{ni} as shown in formula (11.8).

$$\ln \frac{P_{ni}}{1 - P_{ni}} = \beta_0 + \beta_1 X \tag{11.8}$$

Removing the logarithm sign and slightly transforming yields the probability function as shown in formula (11.9).

$$P_{ni} = \frac{e^{\beta_0 + \beta_1 X}}{e + e^{\beta_0 + \beta_1 X}} = \frac{1}{1 + e^{-(\beta_0 + \beta_1 X)}} \tag{11.9}$$

The scenarios discussed involve only two possibilities: occurrence or non-occurrence, hence referred to as the "binary logit" model. Often, there are multiple options to choose from, requiring the use of a multinomial logit model, also known as the MNL model. The multinomial Logit model can be seen as a joint estimation of multiple binary Logit models constructed by pairing each choice behavior in the dependent variable. For rational decision-makers, the probability of choosing option i is as follows, which is also a commonly used discrete choice model. This book adopts this model for the simulation of household and business locations. Assuming the difference of random terms $(\varepsilon_{ni} - \varepsilon_{nj})$ in formula (11.6) follows an independent standard Gumbel extreme value distribution, it can be mathematically proven that formula (11.10) holds true, which is the commonly used discrete choice model (McFadden 1978).

$$P_i = \frac{e^{V_i}}{\sum e^{V_i}} \tag{11.10}$$

11.4.3 Entropy Method

In urban spatial evolution simulation studies, the entropy method is often borrowed to evaluate the weights of indicators.

1. **Principle of Entropy**

Entropy, a concept from thermodynamics, is interpreted in philosophy and statistical physics as the degree of chaos and disorder brought by a material system. Information theory views entropy as the degree of uncertainty of the states of an information source; the greater the amount of information, the smaller the uncertainty and entropy, and vice versa. Based on the characteristics of entropy, we can determine the randomness and disorder level of an event by calculating its entropy value. Entropy can also be used to determine the dispersion of an indicator; the greater the dispersion of the indicator, i.e., the larger the individual differences, the greater the weight of that indicator. The entropy method determines weights based on the concept and properties of entropy, analyzing the weights of various indicators based on their information content or uncertainty.

2. **Calculation Steps**

Assuming the initial data matrix $X = \{x_{ij}\}_{m*n}$ of n indicators for m samples has been obtained, the steps to determine indicator weights by the entropy method are as follows:

First, eliminate the dimensions of indicators. Due to significant differences in the dimensions, magnitudes, and orientations of indicator excellence, it's necessary to perform nondimensionalization on the initial data. The method is chosen based on the actual characteristics and nature of the samples, such as normalization, max–min standardization, zero-mean standardization, etc. The normalized indicator values will be within the [0, 1] interval. Using normalization as an example, the processed matrix is denoted as $Y = \{y_{ij}\}_{m*n}$, with the formula below, where x_{ij} is the value of the Jth indicator for the Ith sample, and $max(x_{ij})$ is the maximum value of the indicator j among the samples.

$$y_{ij} = \frac{x_{ij}}{\max(x_{ij})} \quad j = 0, 1, 2, \ldots. \tag{11.11}$$

Second, determine the entropy of indicators. Entropy represents the turbidity of information, calculated as follows, where y_{ij} is the standardized value of the jth indicator for the ith sample, P_{ij} represents the contribution of the ith sample to the jth indicator, and E_j is the entropy of indicator j. The coefficient $1/ln(m)$ in the entropy formula, where m is the number of samples, ensures that the entropy value ranges from [0,1].

$$p_{ij} = \frac{y_{ij}}{\sum_{i=1}^{m} y_{ij}} \quad j = 1, 2, 3, \ldots \tag{11.12}$$

$$E_j = -\frac{\sum_{j=1}^{n} p_{ij} * \ln(p_{ij})}{\ln(m)} \tag{11.13}$$

Third, determine indicator weights. Mathematically, it can be proven that the greater the entropy E of an indicator, the smaller its weight. Therefore, we further process the entropy values and normalize them to indicator weights w_j.

$$\mathrm{d}_j = 1 - E_j \tag{11.14}$$

$$W_j = \frac{d_j}{\sum_{j=1}^{m} d_j} \tag{11.15}$$

Other methods, such as expert scoring, analytic hierarchy process (AHP), factor analysis, and principal component analysis, can also determine indicator weights, but they are not detailed in this book.

11.4.4 Principal Component Analysis

In empirical research, to analyze problems comprehensively and systematically, we must consider numerous influencing factors, generally referred to as indicators or variables in multivariate statistical analysis. Since each variable reflects certain information about the research question to varying degrees and there is a certain correlation between the indicators, the statistical data obtained overlap to some extent in the information they reflect. When using statistical methods to study multivariate problems, having too many variables increases the computational load and complexity of the problem. However, there is often a desire to involve fewer variables in the quantitative analysis process while obtaining more information. Principal Component Analysis (PCA) is an ideal tool developed to meet this requirement, designed to address such issues.

Principal Component Analysis (PCA), also known as principal component analysis, is a method that involves an appropriate mathematical transformation to generate new variables (i.e., principal components) as linear combinations of the original variables. A few principal components that account for a large proportion of the total information are selected for analysis. The aim of PCA is to use the idea of dimensionality reduction to transform multiple indicators into a few composite indicators, called principal components, with each principal component being a linear combination of the original variables. Moreover, the principal components are uncorrelated with each other, which gives them some superior properties over the original variables. This allows for a focus on a few principal components when studying complex issues without losing too much information, thereby simplifying the problem, revealing the regularity among the internal variables, and improving the efficiency of analysis (Li et al. 1999).

11.4.5 Cobb–Douglas Production Function

The Cobb–Douglas production function is one of the most widely applied utility functions in economics. It was devised by the American mathematician Cobb and the economist Douglas based on data from the American manufacturing sector from 1899 to 1922. Together, they explored the relationship between output and input, incorporating technological resources into the general form of the utility function, and proposed this function form in 1928 (Cobb and Douglas 1928; Goldberger 1968). They suggested that the relationship between output and the inputs of labor and capital, given constant technological levels, can be expressed as:

$$Y = AK^{\alpha}L^{\beta}\lambda \tag{11.16}$$

Here, Y represents the output, A is the technology coefficient, K represents the capital input, L is the labor input; the exponent α is the capital elasticity coefficient, indicating that when production capital increases by 1%, output grows by $\alpha\%$ on average; β is the labor elasticity coefficient, indicating that when the labor input for production increases by 1%, output grows by $\beta\%$ on average; λ is a constant, also known as the efficiency parameter, representing factors that can affect output but cannot be solely attributed to either capital or labor.

When $\alpha + \beta = 1$, the utility function exhibits constant returns to scale, and for any λ, there exists $f(\lambda K, \lambda L) = \lambda f(K, L)$; when $\alpha + \beta > 1, f(\lambda K, \lambda L) > \lambda f(K, L)$, indicating increasing returns to scale; when $\alpha + \beta < 1, f(\lambda K, \lambda L) < \lambda f(K, L)$, indicating decreasing returns to scale.

This function is now widely used in utility evaluation.

11.4.6 Exploratory Space–Time Data Analysis

Exploratory Space–Time Data Analysis (ESTDA) comprises a set of spatial data analysis methods, including Moran's I index, LISA (Local Indicators of Spatial Association) relative path analysis, space–time change analysis, and association network analysis (Jing et al. 2022). Among these, Moran's I index is used for global and local spatial autocorrelation analysis, LISA time path analysis describes the spatial local dependence and variability of elements, space–time change analysis describes the characteristics of element changes, and association network analysis visualizes the collaborative evolution process of spatial elements.

1. **Moran's I Index**

Spatial autocorrelation is divided into global and local autocorrelation. Global autocorrelation describes the distribution of an element across the entire study space, used to determine the clustering characteristics of elements in the study area. The global Moran's I index is commonly used to calculate global autocorrelation. The calculation formula is:

$$I = \frac{N}{\sum\limits_{ij} \omega_{ij}} \frac{\sum_i \sum_j \omega_{ij}(x_i - \overline{x})(x_j - \overline{x})}{\sum_i (x_i - \overline{x})^2} \tag{11.17}$$

In the formula, I represents the global Moran's I index, x_i and x_j represent the element distributions of regions i and j respectively, and $\overline{x}$ is the average value of the element distributions across all regions. ω_{ij} is the spatial weight matrix, which describes the adjacency relationship between two samples in space; N is the number of samples. The value of I ranges from $[-1, 1]$. A positive I indicates a positive spatial correlation among neighbors within a region, and the larger the I, the stronger this correlation. A negative I indicates a negative spatial correlation within the region.

Local spatial autocorrelation further reflects the local spatial relationships between units within a region and can be measured using the local Moran's I index, or LISA index. Its formula is:

$$I_i = z_i \sum_j w_{ij} z_j \tag{11.18}$$

In the formula, I_i represents the local Moran's index, z_i and z_j represent the standardized values of the elements for regions i and j respectively, and w_{ij} is the row-standardized contiguity matrix.

2. **LISA Time Path**

In a Moran scatter plot, the coordinate position of each sample element indicates the type of clustering with its neighbors. Connecting these positions in a time series forms a dynamic LISA time path. LISA time paths not only study the spatiotemporal co-evolution of elements in regions but also determine local spatial differences and temporal dynamics, thus achieving a continuous expression of local spatial dependence from "instant scenes" to "interactive dynamic scenes" (Zhang and Pan 2021). The geometric characteristics of the LISA time path mainly include path relative length, curvature, and movement direction. The calculation models for path relative length AA and curvature BB are:

$$\Gamma_i = \frac{\mathrm{n}^* \sum_{i=1}^{\mathrm{T}-1} d\left(L_{i,t}, L_{i,t+1}\right)}{\sum_{i=1}^{\mathrm{n}} \sum_{t=1}^{\mathrm{T}-1} d\left(L_{i,t}, L_{i,t+1}\right)} \tag{11.19}$$

$$\Delta_i = \frac{\sum_{t=1}^{\mathrm{T}-1} d\left(L_{i,t}, L_{i,t+1}\right)}{d\left(L_{i,t}, L_{i,t+1}\right)} \tag{11.20}$$

In the formula, $L_{i,t}$ is the position of region i in the Moran scatterplot for year t; $d(L_{i,t}, L_{i,t+1})$ is the distance moved by region i between years t and $t+1$; n is the number of regions. A time path relative length greater than 1 indicates that the movement of the region is more dynamic; otherwise, it is more stable. The greater the curvature of the time path, the more significant the spatiotemporal dependency effect among the changes in elements between regions.

The movement direction of the LISA time path, calculated based on the change in the position of regions in the Moran scatterplot across different years, reveals the competitive-cooperative stance among neighborhoods. When the movement direction is between 0°–90° and 180°–270°, the change trend of elements in the region and its neighbors is consistent, showing an integrated spatial dynamic. The former represents a positive synergy, where both the region and its neighbors exhibit a high growth trend, which is a "win–win" situation, while the latter represents negative synergy, where both exhibit a low growth trend, indicating a "lose-lose" situation. Movement directions between 90°–180° and 270°–360° indicate opposite trends of change between the region and its neighbors, the former being a "lose-win" situation and the latter a "win-lose" situation.

1. **LISA Space–Time Transitions**

LISA space–time transitions are used to describe the characteristics of the spatial relationships among neighborhood elements over time. Through an improved Markov chain to measure sample transitions, four types and sixteen different space–time transitions can be defined (Rey and Ye 2010). The specific types are categorized as shown in Table 11.1.

HH (first quadrant) represents high-high clustering, LH (second quadrant) low–high clustering, LL (third quadrant) low-low clustering, and HL (fourth quadrant) high-low clustering. Transitions between HL and LH are referred to as heterogeneity changes, while transitions between HH and LL are called homogeneity changes. The Spatial Homogeneity Trend Indicator (SHTI) is defined as:

$$\text{SHIT} = \frac{H_m - H_t}{2} \tag{11.21}$$

H_m and H_t respectively represent the probability of transitioning from heterogeneity to homogeneity and homogeneity to heterogeneity. The range of SHTI values is $[-1, 1]$, with positive values indicating a tendency towards spatial homogeneity in transitions, and negative values indicating the opposite.

Table 11.1 Types of space–time transitions

Type	Description	Transition direction
I	Self transition, neighborhood stable	$HH_t \rightarrow LH_{t+1}$, $LH_{t+1} \rightarrow HH_{t+1}$, $HL_t \rightarrow LL_{t+1}$, $LL_t \rightarrow HL_{t+1}$
II	Self stable, neighborhood transition	$HH_t \rightarrow HL_{t+1}$, $LH_t \rightarrow LL_{t+1}$, $HL_t \rightarrow HH_{t+1}$, $LL_{t\rightarrow} LH_{t+1}$
III	Both self and neighborhood transition	$HH_t \rightarrow LL_{t+1}$, $LH_t \rightarrow HL_{t+1}$, $HL_t \rightarrow LH_{t+1}$, $LL_t \rightarrow HH_{t+1}$
IV	Both self and neighborhood stable	$HH_t \rightarrow HH_{t+1}$, $LH_t \rightarrow LH_{t+1}$, $HL_t \rightarrow HL_{t+1}$, $LL_t \rightarrow LL_{t+1}$

11.4.7 Multiscale Geographically Weighted Regression

Multiscale Geographically Weighted Regression (MGWR) is a type of local regression that considers the impact of spatial relationships on the model, expanding upon the classic geographically weighted regression by adding different spatial observation scales for multivariate data and calculating the optimal bandwidth for each variable. This approach not only captures the spatial heterogeneity in regression relationships but also predicts trends in spatial clustering of variables (Fotheringham et al. 2017). The MGWR calculation model is:

$$y_i = \sum_{1}^{k} \beta_{b\omega j}(\mu_i, v_i)x_{ij} + \varepsilon_i \tag{11.22}$$

In the MGWR model, y_i represents the dependent variable, which is the element to be tested. x_{ij} represents the impact factor j of region i, and $\beta_{b\omega j}(\mu_i, v_i)$ represents a continuous function of the spatial position (μ_i, v_i) of region i. ε_i denotes independent random errors. $\beta_{b\omega j}(\mu_i, v_i)$ is calculated through the spatial weight matrix as follows:

$$\beta_{b\omega j}(\mu_i, v_i) = \left(X_j' W_{b\omega j}(\mu_i, v_i)X_j\right)^{-1} X_j W_{b\omega j}(\mu_i, v_i)y \tag{11.23}$$

where $W_{b\omega j}(\mu_i, v_i)$ represents the spatial weights, and y represents the observed outcomes. In the MGWR model, each explanatory variable has its own spatial scale, and the optimal bandwidth for variables is detected using the corrected Akaike Information Criterion (AICc). When the bandwidth decreases over time, the variation in the dependent variable y_i tends to show a tendency towards spatial clustering.

References

Bivand RS, Pebesma EJ, Gomez-Rubio V (2013) Applied spatial data analysis with R. Aiping X, Hong S (trans). Tsinghua University Press, Beijing (in Chinese)

Cobb CW, Douglas PH (1928) A theory of production. Am Econ Rev 18(1):139–165

Fotheringham AS, Yang W, Kang W (2017) Multiscale geographically weighted regression (MGWR). Ann Am Assoc Geogr 1247–1265

Goldberger AS (1968) The interpretation and estimation of Cobb-Douglas functions. Econometrica 35(3/4):464–472

Jing S, Yan Y, Niu F, et al (2022) Urban expansion in China: spatiotemporal dynamics and determinants. Land 11

Kabacoff RI (2016) In: Xiaoning W et al (trans) R in action, 2nd edn. Posts & Telecom Press, Beijing (in Chinese)

Li Y, Zeng Z, Zhang M, et al (1999) Application of primary component analysis in methods of comprehensive evaluation for many indexes. J Hebei Univ Technol 1:96–99 (in Chinese)

Long Y (2021) Principles and applications of urban models. China Architecture & Building Press, Beijing (in Chinese)

McFadden D (1978) Modelling the choice of residential location. In: Karlquist A et al (eds) Spatial interaction theory and residential location. North Holland, Amsterdam, pp 75–96
McKinney W (2018) In: Jingyi X (trans) Python for data analysis. Machinery Industry Press, Beijing (in Chinese)
Microsoft Excel (2018). https://www.microsoft.com/zh-cn/microsoft-365/excel/
Netlogo. http://ccl.northwestern.edu/netlogo/download.shtml
Nie C, Jia S (2005) Resarch on the theoretical basis and evolution of discrete choice models. J Quant Tech Econ 11:151–159 (in Chinese)
Rey SJ, Ye X (2010) Comparative spatial dynamics of regional systems. In: Advances in spatial science
Shi H (2015) Factors affecting urban residents' low-carbon consumption behavior in Fuzhou. Resour Sci 37(2):308–317 (in Chinese)
Wang C, Wang D, Zhu W, et al (2015) Research progress of discrete choice models. Prog Geogr 34(10):1275–1287 (in Chinese)
Wu J, Cao Y, Yao S, et al (2009) Study on the evaluation of coordinated development of infrastructure and regional economic system. Econ Geogr 29(10):1624–1628 (in Chinese)
Yan H (2018) Consumption preference analysis based on discrete choice model. Chengdu University of Technology (in Chinese)
Zhang X, Pan J (2021) Spatio-temporal dynamic identification and driver detection of urban sprawl in China. Hum. Geogr 36:114-125

Epilogue

Simulation is the fitting of the development process of things. The premise of undertaking simulation work is to have a full understanding and recognition of the operational process of things and to express it abstractly. For example, the simulation of the urban spatial evolution process requires a clear understanding of the composition of urban space, a systematic recognition of the interaction relationships between various components, and the operational process of urban space. Although the ultimate realization of simulation work is implemented in mathematical formulas or computer algorithms, and the same model can be implemented with different mathematical formulas, mere mathematical formulas cannot be considered a model (simulation), at least not accurately. Compared to empirical research, simulation work should be said to be deductive, an application of the deduction of the laws of things. Therefore, before modeling, an in-depth analysis is required, and a full grasp of the development laws of things is needed.

Since the development of things is usually a complex process, research work is hard to exhaust all influencing factors and their mechanisms of action. Therefore, simulation work focuses on capturing the main influencing factors. The simulation results cannot accurately represent the future, especially for development processes that involve human or policy interventions, such as urban and regional development processes, or what is often said, "all models are wrong." However, models can support decision-making by comparing simulation results under different scenarios. For example, under the same conditions, the implementation of a certain policy and the scenario of not implementing the policy will yield different simulation results, and the difference can be regarded as the impact of that policy. Therefore, in decision support, the relative magnitude of simulation values is more meaningful than the absolute values of simulation results.

The urban spatial evolution simulation model is an integrated model. The modeling of the model is not about innovating each module (sub-model) but about the integrated application of various modules. This book discusses a large number of models involved in the construction process of urban models. Some models are original, and some are proposed by predecessors. If any references are omitted, please contact the author. It is hoped that this book can provide some help to readers,

F. Niu, *Urban Spatial Evolution Simulation*, Urban Sustainability,
https://doi.org/10.1007/978-981-97-3481-8

especially those from the field of human geography, in learning and mastering "simulation" methods.

Empirical research has always been prevalent in China's human geography field. Currently, quantitative research is gradually rising and gaining more attention. Quantitative research is irreplaceable in scenario testing. As quantitative research continues to develop, establishing a simulation economic geography disciplinary system and enriching the content of human geography has significant importance. There are many areas in this book that need to be detailed, and this is an attempt to inspire further discussion. The author will continue to explore and further refine the model. It is hoped that more scholars will join in the effort to advance the establishment of a simulation economic geography disciplinary system.